AF699694

Agrochemical Resistance

About the Cover

Extent, selection, and diagnosis or resistant insects (top), weeds (middle), and pathogenic fungi (bottom) at increasing levels of selection pressure (left to right) by insecticides, herbicides, and fungicides, respectively. Resistant organisms are highlighted. The bottom row illustrates the detection of resistant organisms by biochemical assay, such as ELISA.

ACS SYMPOSIUM SERIES **808**

Agrochemical Resistance

Extent, Mechanism, and Detection

ARTHUR LAKES LIBRARY
COLORADO SCHOOL OF MINES
GOLDEN, CO 80401

J. Marshall Clark, Editor
University of Massachusetts

Isamu Yamaguchi, Editor
Riken Institute

JAN 2002
RECEIVED
Colorado School of Mines Library

American Chemical Society, Washington, DC

SB
957
.A45
2002

Library of Congress Cataloging-in-Publication Data

Agrochemical resistance: extent, mechanism, and detection / J. Marshall Clark, editor, Isamu Yamaguchi, editor.

p. cm.—(ACS symposium series ; 808)

Includes bibliographical references and index.

ISBN 0–8412–3723–9

1. Pesticide resistance—Congresses.

I. Clark, J. Marshall (John Marshall), 1949 - II. Yamaguchi, Isamu, 1941- III. Pan Pacific Conference on Pesticide Science (2nd ; 1999 ; Honolulu. Hawaii) IV. Series.

SB957 .A45 2001
632'.95042—dc21 2001045700

The paper used in this publication meets the minimum requirements of American National Standard for Information Sciences—Permanence of Paper for Printed Library Materials, ANSI Z39.48–1984.

Copyright © 2002 American Chemical Society

Distributed by Oxford University Press

All Rights Reserved. Reprographic copying beyond that permitted by Sections 107 or 108 of the U.S. Copyright Act is allowed for internal use only, provided that a per-chapter fee of $20.50 plus $0.75 per page is paid to the Copyright Clearance Center, Inc., 222 Rosewood Drive, Danvers, MA 01923, USA. Republication or reproduction for sale of pages in this book is permitted only under license from ACS. Direct these and other permission requests to ACS Copyright Office, Publications Division, 1155 16th St., N.W., Washington, DC 20036.

The citation of trade names and/or names of manufacturers in this publication is not to be construed as an endorsement or as approval by ACS of the commercial products or services referenced herein; nor should the mere reference herein to any drawing, specification, chemical process, or other data be regarded as a license or as a conveyance of any right or permission to the holder, reader, or any other person or corporation, to manufacture, reproduce, use, or sell any patented invention or copyrighted work that may in any way be related thereto. Registered names, trademarks, etc., used in this publication, even without specific indication thereof, are not to be considered unprotected by law.

PRINTED IN THE UNITED STATES OF AMERICA

Foreword

The ACS Symposium Series was first published in 1974 to provide a mechanism for publishing symposia quickly in book form. The purpose of the series is to publish timely, comprehensive books developed from ACS sponsored symposia based on current scientific research. Occasionally, books are developed from symposia sponsored by other organizations when the topic is of keen interest to the chemistry audience.

Before agreeing to publish a book, the proposed table of contents is reviewed for appropriate and comprehensive coverage and for interest to the audience. Some papers may be excluded to better focus the book; others may be added to provide comprehensiveness. When appropriate, overview or introductory chapters are added. Drafts of chapters are peer-reviewed prior to final acceptance or rejection, and manuscripts are prepared in camera-ready format.

As a rule, only original research papers and original review papers are included in the volumes. Verbatim reproductions of previously published papers are not accepted.

ACS Books Department

Contents

Insecticide Resistance

Herbicide Resistance

Fungicide Resistance

Preface

Pesticide science has been given the important task of providing human society the food, fiber, and health that it requires in an environmentally safe, sustainable, and affordable manner. One of the main challenges that we face in obtaining these goals is pesticide resistance. It is well established that the genetic-based evolutionary selection processes that organisms use to develop chemical-resistant strains are universal throughout biology. It is now obvious that resistance to pesticides, which extends to bactericides, fungicides, herbicides, insecticides, nematicides, and rodenticides, will occur for any pest control product when large populations are treated repeatedly with no other intervention. Today, the added pesticide costs associated with resistance is estimated in the millions of dollars and in the billions of dollars when the costs of lost yields are incorporated.

Given the genetic basis of pesticide resistance in insects, weeds, and plant pathogens, it is not surprising that we have adopted a genetic approach to its study. Molecular biology approaches have changed the way in which we develop new and novel-acting plant protection products, determine mechanisms of resistance, monitor for resistant pest organisms and produce transgenic crops that are phenotypically less desirable as hosts. How long these new products and crop varieties remain as effective crop protection strategies will depend in large part on how well we manage them in terms of resistance development.

Pesticide resistance management has become a primary consideration for any newly introduced crop protection product and an essential component for integrated pest management (IPM) strategies. The ultimate goal of resistance management is the preservation of susceptible genes within the treated pest population. Thus, a major approach in any control strategy is to use all other IPM tools that are available prior to the application of synthetic pesticides. Nevertheless, pesticides, old and new, still remain our most widely used, most effective and most irreplaceable means to control pest organisms. To effectively manage the application of pesticides, diagnostic means to monitor the frequency of resistant alleles in pest organisms is fundamental and a critical need. Such diagnostic techniques must be cheap, rapid, rugged, and of high enough resolution to identify heterozygote individuals. Clearly, DNA-based diagnostic procedures have many of these attributes and their development is rapidly expanding.

In response to such concerns, the American Chemical Society (ACS) Division of Agrochemicals and the Pesticide Science Society of Japan jointly convened the 2nd Pan-Pacific Conference on Pesticide Science on October 24–27, 1999, in Honolulu, Hawaii. Scientists from 12 nations, primarily associated with the Pacific-Rim area, met to discuss the issues of pesticide resistance and management and the papers presented are the nucleus for this book.

The book is divided into three main sections; insecticide resistance, herbicide resistance, and fungicide resistance. Each section covers topics on the occurrence and extent of resistance, molecular determination of specific resistance mechanisms and diagnostic techniques for resistance management.

Acknowledgments

We thank all the authors for their presentations at the Conference and for their contributed chapters. A special thanks to our organizers of the three subtopics: insecticide resistance (Fumio Matsumura, University of California, Davis) and Gary Fitt (CSIRO, Australia); herbicide resistance (Kriton Hatzios, Virginia Tech) and Yasuhiro Yogo (NARC, Japan); and fungicide resistance Wolfram Koeller (Cornell University) and Hideo Ishii (NIAES, Japan). In particular, we extend our deepest appreciation to the many expert colleagues who provided helpful and necessary critical reviews. We thank Anne Wilson, Kelly Dennis and Stacy VanDerWall of the ACS Books Department for all their help, suggestions, and encouragement. Lastly, we thank the ACS Division of Agrochemicals and their benefactors, contributors, and donors who's financial support made this book possible.

J. Marshall Clark
Department of Entomology
University of Massachusetts
Amherst, MA 01003

Isamu Yamaguchi
Bio-Environmental Science
Riken Institute
2–1 Hirosawa, Wako
Saitama 351–0198
Japan

Chapter 1

Scope and Status of Pesticide Resistance

J. Marshall Clark[1] and Isamu Yamaguchi[2]

[1]Department of Entomology, University of Massachusetts, Amherst, MA 01003

[2]Bio-Environmental Science, Riken Institute, 2–1 Hirosawa, Wako, Saitama 351–0198, Japan

The role of pesticide science is to provide society the food, fiber and health that it requires in an environmentally safe, sustainable and affordable manner. Resistance to the crop protection products that we use to control insect pests, weeds and plant diseases is a major problem we face today in successfully attaining these goals. Growing crops today requires the combined use of novel crop protection products, transgenic plants and integrated management techniques. New chemistry is expensive with the cost of the development, registration and marketing of a single new product estimated at $75-150 million over a period of 8-10 years. Resistance to our widely used pesticides is common, extensive and increasing. To protect the susceptibility within targeted populations and our initial research and development investment, pesticide resistance management must be practiced whenever crop protection products are used. This

© 2002 American Chemical Society

chapter will summarize our current understanding of resistance to insecticides, herbicides and fungicides and compare their similarities and differences. The current status of resistance, means to manage resistance and the estimation of resistance "risk" will be discussed for each pesticide group.

The evolutionary selection processes that organisms use to develop chemical-resistant strains are universal throughout biology. It should be evident that, although first identified for insecticides, resistance to pesticides now extends to bactericides, fungicides, herbicides, nematicides and rodenticides and will occur for any pest control product when large populations are treated repeatedly with no other intervention. The following summarization of the scope and status of resistance to insecticides, herbicides and fungicides is taken largely from the information provided by the Insecticide Resistance Action Committee (IRAC), the Herbicide Resistance Action Committee (HRAC) and the Fungicide Resistance Action Committee (FRAC), respectively (1-3).

Insecticide Resistance

Background

Resistance to insecticides has been defined as "the developed ability in a strain of insects to tolerate doses of toxicants which would prove lethal to the majority of individuals in a normal (previously untreated) population of the same species" (4). Simply put, insecticide treatments remove susceptible individuals from the population and allow only those individuals that possess resistance genes to reproduce. Resistance genes most likely already exist in most insect population due to polymorphism or may occur by mutation. If treatment is continued, eventually resistant insects will replace susceptible individuals and the insecticide will become increasingly less effective as a crop protection agent. The rate at which resistance evolves is dependent upon the selection pressure placed on the population and is controlled by three aspects; the biology of the pest insect, the chemistry and specificity of the insecticide and the treatment regime used. Insects that have high reproductive rates, low migration behavior and a restricted host range tend to become resistant more quickly that those without these traits. Chemicals that have long environmental persistence and a single mode of action have higher resistance potential than those that are short lived and have multiple modes of action. Resistance likewise occurs more rapidly when the insecticide is used more intensely (high dose rate and increased number of application per growing season).

The first documented cases of insecticide resistance were reported by Melander in 1914 in the San Jose scale to lime sulfur and by Quayle in 1916 in the California red scale to cyanide (5). By 1945, 13 insect or tick species were determined to be resistant to arsenicals, selenium, rotenone, cyanide and various other inorganic insecticides. Since the introduction of DDT in 1942 and its first report of resistance in the house fly in 1947, resistance to synthetic organic insecticides have been reported some 2 to 20 years later at an ever increasing level with each introduced chemical class (6), including the cyclodienes, carbamates, formamidines, organophosphates, pyrethroids, avermectins, chloronicotinyls, *Bacillus thuringiensis*, etc. Today, the added insecticide costs associated with resistance in the USA alone is estimated at $40 million and greater that $1 billion when the costs of lost yields are incorporated (1).

Mechanisms

Insects and mites use a variety of means in becoming resistant to insecticides including; xenobiotic metabolism, target site insensitivity, altered toxicokinetics and behavior. Metabolic resistance is the most common resistance mechanism and results in the rapid and efficient enzymatic detoxification of the insecticide. Resistant insects either produce more enzyme, more efficient forms of the enzyme or both. The most common enzymes involved are the cytochrome P_{450} monooxygenases, hydrolytic esterases and glutathione *S*-transferases. Target site resistance results from an alteration in the structure of the insecticide receptor molecule, which reduces the binding of the insecticide to its receptor, and results in the insensitivity of the insect to the insecticide. Most common target site alterations leading to resistance are GABA-chloride channels to cyclodienes and related compounds, voltage-sensitive sodium channels to DDT and the pyrethroids, and acetylcholinesterases to organophosphorous and carbamate insecticides. Alterations in the toxicokinetics of insecticides, such as reduced penetration/absorption, sequestration and enhanced excretion rates, have been determined in a wide range of resistant insects to a variety of insecticides. Although not as effective as the above two mechanisms in producing high levels of resistance themselves, they do interact with many other mechanisms in a synergistic fashion and can result in greatly enhanced levels of resistance.

Behavioral resistance (e.g., avoidance or reduced feeding behaviors) has been determined in insects and mites to a variety of insecticides/acaricides including; the organochlorines, organophosphates, carbamates and pyrethroids.

Occurrence and Extent

There are currently at least 540 arthropod species that are resistant to one or more insecticides and acaricides. These arthropods are resistant to 310 different insecticides, acaricides or formulations. These resistant insect species represent 15 different arthropod orders and includes the acari, araneae, coleoptera, dermaptera, diptera, ephemeraptera, hemiptera, homoptera, hymenoptera, lepidoptera, neuroptera, orthoptera, phthiraptera, siphonaptera and thysanoptera. As indicated in Table I, the top five orders comprise approximately 90% of the resistant species. The flies and mosquitoes (diptera) are the most numerous, followed by the mites and ticks (acari), butterflies and moths (lepidoptera), beetles (coleoptera) and the homoptera, including the aphids, cicadas, hoppers, psyllids, scales and whiteflies. Table II lists those arthropod species that have developed resistance to insecticides/acaricides most frequently. Seven of the top 10 species to show resistance are agricultural pests and 3 are medically important. Resistance has been found most often in the two-spotted spider mite in agricultural situations and in the southern cattle tick for non-agricultural situations.

There are representative insecticides/acaricides to which arthropods have become resistant in 14 major pesticide classes including the avermectins, *Bacillus sp.*, carbamates, formamidines, fumigants, inorganics, insect growth regulators (IGR), organochlorides, organophosphates, organotins, neonicotinyls, pyrazoles, pyrethroids and rotenoids. Of the 22 modes of actions listed in the IRAC Mode of Action Classification (1), there are insects resistant to insecticides/acaricides that act through at least 9 separate mechanisms including acetylcholinesterase inhibitors, GABA-gated chloride channel antagonists, sodium channel agonists, acetylcholine receptor modulators, chloride channel activators, juvenile hormone mimics, microbial disrupters of insect midgut membranes, oxidative phosphorylation inhibitors, and site I electron transport inhibitors. Resistance to the organochloride pesticides is most common and comprises approximately 66% of the top 10 insecticides/acaricides or formulations that have a arthropod species resistant to them (Table III). Organophosphorous products make up an additional 29% and the carbamates are 5% of this total.

Table I. Most resistant arthropod orders to insecticides or acaricides[a]

Order	Number of resistant species	Percent of total number of resistant species (548)
Diptera	184	34%
Acari	83	15%
Lepidoptera	82	15%
Coleoptera	73	13%
Homoptera	58	11%
	480	88%

[a] Data are from reference 1.

Table II. Most resistant arthropod species to insecticides or acaricides[a]

Species	Common name	Impact area	Host/economic impact	Number of pesticide
Tetranychus urticae	two-spotted spider mite	agric[b]	cotton, fruits ornamentals walnuts	72
Plutella xylostella	diamond-back moth	agric	crucifer, nasturtium	69
Myzus persicae	peach-potato aphid	agric	crops, flower, fruit, grains, trees, tobacco, vegetables	68
Boophilus microplus	southern catle tick	med[c]	cattle	41
Leptinotarsa decemlineata	Colorado potato beetle	agric	eggplant, pepper potato, tomato	40
Panonychus ulmi	red spider mite	agric	fruit, trees	40
Blattella germanica	German cockroach	med	urban	39
Heliothis virescens	tobacco bud worm	agric	chickpea, corn cotton, tobacco	37
Musca domestica	house fly	med	urban	36
Tribolium castaneum	red flour beetle	agric	peanuts, sorghum stored grain	34

[a] Data are from reference 1.

[b] Agric = agricultural.

[c] Med = medical.

Table III. Insecticides, acaricides or formulations that have the most arthropod species resistant to them[a]

Insecticide, acaricide or formulation	Pesticide class	Total number of resistant species
DDT	organochloride	257
BHC/cyclodienes	organochloride	149
malathion	organophosphorous	120
dieldrin	organochloride	100
aldrin	organochloride	87
lindane/BHC	organochloride	76
parathion	organophosphorous	76
diazinon	organophosphorous	58
carbaryl	carbamate	51
fenitrothion	organophosphorous	47

[a] Data are from reference 1.

Management

Insecticides and acaricides are still the main tools used to control arthropod pests and preserve yields. Along with a variety of effective older products, transgenic crops and IPM strategies, a number of exciting new compounds have recently been introduced including; fiproles (fipronil), chloronicotinyls (imidacloprid), spinosyns (Spinosad), non-steroidal ecdysone agonists (tebufenozide), oxadiazine sodium channel antagonists (indoxacarb), etc. These new crop protection products act at novel insect target sites and have elicited little cross-resistance to previously used pesticides. As such, they have been shown to be extremely effective and, most importantly, amenable to insecticide resistance management strategies.

Insecticide resistance management is an important component of IPM and a primary consideration for any newly introduced crop protection product. The ultimate goal of resistance management is the preservation of susceptible genes within the treated insect pest population. This process is accomplished by three basic steps; monitoring the development and density of pest populations, establishment of economic threshold levels of injury and the implementation of integrated control strategies. The last step is perhaps the most useful and most amenable to manipulation. A major consideration in any control strategy is to use all IPM tools that are available including; cultural practices, crop rotations, pest-resistant crop varieties, transgenic plants, beneficial insects, chemical attractants or deterrents prior to the application of biological or synthetic insecticides. Applications of any crop protection product should follow the manufacturer's instructions and be made against the most vulnerable life stage of the insect pest. When synthetic insecticides must be used, they should be rotated between product class and modes of action.

Herbicide Resistance

Background

Resistance of weeds to herbicides has been defined as "the naturally occurring inheritability of some weed biotypes within a given weed population to survive an herbicide treatment that should, under normal use conditions, effectively control that weed population" (2). The assumption is that in any weed population there are a small number of plants resistant to a given herbicide and upon repeated applications will survive and set seed. The resistant biotype will become the dominant form in the population over a period of time following repeated applications. The rate at which resistance evolves is dependent upon the selection pressure placed on the plant population and is controlled by three aspects; the biology and genetics of the weed species, the chemistry and specificity of the herbicide and the control measures utilized. Generally, weeds that have high density due to high seed production and low dispersal behavior tend to become resistant more quickly than those without these traits. Weeds with low genetic diversity and short soil dormancy for their seed prior to germination will have a higher propensity towards developing resistance. The higher the fitness of the resistant plant phenotype relative to the susceptible phenotype, the faster resistance will evolve. Herbicides that are more environmentally persistent have a common mode of action with many other herbicides and are more frequently used will evolve resistance more rapidly than compounds without these characteristics. Weed resistance to herbicides will evolve the slowest in cropping situation where full rotations are used and mechanical and cultural controls (cultivation, stubble burning, competitive crops, stale seedbeds, etc) are implemented.

Herbicide resistance was first documented in 1964 when *Senecio vulgaris* (common groundsel) was determined to be resistant to triazine herbicides. By 1994, global herbicide sales were ~\$13 billion and accounted for nearly half of all agrochemical used. Over 65% of the herbicide sold was to North America (42%) and Western Europe (23.2%). However, East Asia (19.1%) and Latin America (9.4%) accounted for ~ 30% and this market has currently shown significant growth. Five crops; maize (18.2%), cereals (16.5%), soybean (14.8%), fruits and vegetables (13.2%) and rice (10.3%), accounted for ~75% of global herbicide use. With increased use, 60 herbicide-resistant plant biotypes had been determined by 1983 and elicited resistance to 6 major herbicide groupings based on common mode of action (Table IV). The number of herbicide-resistant plant biotypes increased to 171 by 1997 and had representative grasses and broadleaf weed biotypes resistant to all 7 major herbicide groupings plus 8 additional "new" mode of action groupings (8,9). In sharp contrast to the situation with insecticides, no herbicides have been lost due to resistance and only under rare circumstances has herbicide resistance become a limiting factor for crop production (e.g., resistant *Lolium rigidum* in Australia and resistant *Alopecurus myosuroides* in Europe).

Mechanisms

Similar to insects, plants use a number of mechanisms to overcome the toxic effects of herbicides. The major mechanisms include; altered target sites, xenobiotic metabolism and enhanced sequestration and compartmentalism.

Table IV. Development of herbicide resistant weeds (1983-1997)[a]

Herbicide group[b]	Number of resistant plant biotypes	
	1983	1997
Triazines	40	60
Synthetic auxins	8	10
Bipyridiums	6	17
ACCase inhibitors[c]	2	19
Dinitroanilines	2	5
Ureas & Amides	1	15
ALS inhibitors[d]	0	63
Others	1	21
	60	210

[a] Data are from reference 2.

[b] Herbicide classification according to primary sites of action, HRAC, 2001.

[c] Inhibitors of acetyl CoA carboxylase (ACCase).

[d] Inhibitors of acetolactate synthase (ALS).

Mutations within structural genes produce altered target proteins that do not bind the herbicide well, leading to target site insensitivity and resistance. Altered target site mechanisms have resulted in resistance to herbicides that inhibit acetolactate synthase (ALS inhibitors) such as the sulfonylureas and inhibitors of photosynthesis at photosystem II such as the triazines. Elevated levels of enzymes involved in xenobiotic metabolism or more efficient metabolism can degrade herbicides to non-phytotoxic metabolites faster in resistant plants than in the susceptible form. *Alopecurus myosuroides* (blackgrass) and *Lolium rigidum* (rigid ryegrass) have both displayed this type of resistance to aryloxyphenoxypropionate herbicides such as clodinafop-propargyl. Enhance sequestration occurs when an endogenous molecule (often a sugar moiety) binds the herbicide, which effectively removes it from any mass action with its target site. Compartmentalization results in the removal of the herbicide from the metabolically active regions of the plant cell, usually into a vacuole. There are a number of plant species that have become resistant to aryloxyphenoxypropionate herbicides and paraquat by these means.

Occurrence and Extent

There are currently 249 plant biotypes encompassing 153 plant species (91 dicots and 62 monocots) resistant to herbicides that have been identified in over 200,000 fields (Table V). The resistant weed biotypes have representatives that are resistant to herbicides classified in 15 of the 24 currently approved HRAC/Weed Science Society of America (WSSA) group classification by mode of action (B, C1, A, D, C2, O K1, N, F3, G, K3, C3, F1, K2 and Z). Herbicide resistance is most pronounced against the ALS inhibitor- and triazine-type herbicides. There are moderate numbers of resistant weeds to the bipyridiliums, ureas, amides, synthetic auxins and the dinitroanilines. There are limited instances of resistant weeds to the thiocarbamates, triazoles, ureas, isoxazolidiones, glycines, chloroactetamides, nitriles, carotenoid biosynthesis inhibitors, mitosis inhibitors, organoarsenicals, arylaminopropionic acids and pyrazoliums. Weed biotypes resistant to the top three HRAC herbicide groups (ALS inhibitors, triazines and the ACCase inhibitors) comprise over 63% of the total number of resistant biotypes. Table VI lists the most common weed species by occurrence (more than 10 independent reports) for each herbicide group. *Chenopodium album* (lambsquarters) is the most common resistant weed (to triazines), followed by *Amaranthus retroflexus* (redroot pigweed, to traizines), *Avena fatua* (wild oat, to ACCase inhibitors) and *Kochia scoparia* (kochia, to ALS inhibitors). Of the Pacific-Rim countries, the Untied States of America has the highest number of resistant weed biotypes (85), followed by Australia (36), Canada (35), Japan (14), Malaysia (12), Chile and New Zealand (7 each), China (5), Costa Rica (4), South Korea (3), Mexico and Thailand (2 each), and Columbia, Ecuador, Fiji, Indonesia, Philippines, and Taiwan (1 each).

Table V. Current status of herbicide resistance[a]

Herbicide group (HRAC code)	site of action	Group example	Number of resistant biotypes
ALS inhibitors (B)	acetolactate synthase	prosulfuron	69
Triazines (C1)	photosystem II	simazine	63
ACCase inhibitors (A)	acetyl CoA carboxylase	cyclofop-butyl	25
Bipyridiliums (D)	photosystem I	diquat	21
Ureas & Amides (C2)	photosystem II	diuron	20
Synthetic auxins (O)	indoleacetic acid receptor	mecoprop	20
Dinitroanilines (K1)	tubulin assembly	benefin	10
Others			21
			249

[a] Data are from reference 2.

Table VI. Most common weed species resistant to each of the three most impacted HRAC herbicide groups[a]

Herbicide group (HRAC code)	Weed species	Common name	Year	Number of reported occurrences
ALS inhibitors (B)	*Kochia scoparia*	Kochia	1987	21
	Avena fatua	wild oat	1986	11
Triazines (C1)	*Chenopodium album*	lambsquarters	1973	37
	Amaranthus retroflexus	redroot pigweed	1980	22
	Amaranthus hybridus	smooth pigweed	1972	16
	Senecio vulgaris	common grandsel	1970	13
	Poa annua	annual bluegrass	1978	12
	Kochia scoparia	Kochia	1976	11
ACCase inhibitors (A)	*Avena fatua*	wild oat	1985	22
	Lolium multiflorum	Italian ryegrass	1987	12
	Lolium rigidum	rigid ryegrass	1982	12

[a] Data are from reference 2 .

[b] Only reported occurrences above 10 are reported.

Management

Integrated weed management (IWM) has been defined as the use of a range of control measures, including physical, chemical and biological methods, in an integrated fashion without excessive reliance on any single aspect (3). Herbicide resistance occurs usually following intensive selection pressure on a weed population over a number of growing seasons. This process is hastened by repeated use of the same herbicide, using herbicides with the same mode of action, cropping only as monocultures and using little or no cultivation practices. Management practices suggested by the HRAC include the judicial selection of herbicides, applications as mixtures or in rotational schemes, using herbicides with various modes of action, crop rotations and selective cultural practices to minimize the need for chemical intervention. The use of herbicides that have different modes of action in mixtures or rotations can be an effective strategy to delay resistance because the odds of an individual weed having multiple resistance mechanisms simultaneously are low. Often, mixtures are necessary to control the entire weed spectrum regardless. However, mixture components should have similar efficacy against the target weed, have similar biological persistence and be detoxified by different metabolic pathways. Rotation of crops avoids situations where herbicides with the same modes of action and spectrum of efficacy are used to control the same weed flora. Different herbicides and cultivation methods can be implemented and should provide a competitive environment to shift the weed flora. Cultural or non-chemical weed control practices can be very effective in reducing the soil seed bank and reduce the need for chemical intervention. Practices such as ploughing, hoeing, stubble burning, grazing and stale seedbeds have proven effective in reducing weed populations under specific conditions.

Fungicide Resistance

Background

Pathogen resistance to fungicides is widespread and the performance of most modern, systemic fungicides has been affected. Resistance initiates through the survival and reproduction of rare pathogenic mutants that are not controlled during initial exposure to a fungicide. Resistance development occurs when the same fungicide or fungicides that have the same mode of action are repeatedly applied without additional non-chemical intervention. Development is discrete if due to a single gene mutation or gradual if multiple factors are implicated (polygenetic) and can be determined by genetic recombination tests. The rate of resistance evolution is dependent on the selection pressure and occurs most rapidly when the following conditions exist; use of repetitive or sustained treatments, more effective application method or dose, application to large areas, large initial pathogen populations, rapid reproduction of the target fungi, short generation time of the pathogen, cross-resistance to existing fungicides and rapid generation of fit, resistant mutants in the laboratory.

Fungicides, in some fashion, have been used to protect plants from fungal attack for the past 200 years. Fungicide resistance has been determined for the past 25 years, particularly since the introduction of the modern systemic fungicides. Today, approximately 135 different fungicides are produced and sold worldwide. In 1993, the total value of sales to end-users was ~$4.7 billion, with roughly half this amount used in Europe. Initially, copper-based formulations, organo-mercury and sulfur products provided effective treatments and have been extensively used. Interestingly, no resistance to these products has been reported over many years of extensive use. The next generation of fungicides includes the phthalimides, dithiocarbamates, dinitrophenols and the chlorophenyls, which have been used for the past 30 years. During the 1960s and 1970s, the "systemic" fungicides were introduced including; the 2-aminopyrimidines, benzimidazoles, carboxanilides, phosphorothiolates, morpholines, dicarboximides, phenylamides and the sterol demethylation inhibitors (DMIs). More recently, a number of newer fungicides, exploiting novel chemistries, have been developed and include the phenylpyrroles, anilinopyrimidines and the strobilurins. Using estimates of inherent risk of resistance, the FRAC has grouped fungicide or classes of fungicides into resistance risk groups using a variety of criteria. The benzimdazoles, dicarboximides, and phenylamides are in the *high resistance risk group*. The 2-aminopyrimidines, anilinopyrimidines, aromatic hydrocarbons, azoles, carboxanilides, cymoxanil, dimethomorph, fentins, phenylpyrroles, phosphorothiolates, pyrimidinecarbinols and strobilurins are in the *moderate resistance risk group*. The *low resistance risk group* consists of acibenzolar-S-methyl, chlorothalonil, copper-based compounds, dithiocarbamates, fluazinam, phthalimides, probenazole, quinoxyfen, sulfurs and tricyclazole.

Mechanisms

In common with insects and plants, fungi may employ a number of mechanisms, either singularly or in combination, to become resistant to fungicides. The major mechanisms include; alteration of the target receptor so that it becomes insensitive, enhanced xenobiotic metabolism that detoxifies the fungicide, alternative biochemical pathway that avoids the target receptor and exclusion or expulsion of the fungicide by the fungal cell. Table VII presents the current information on the mechanisms of resistance to various fungicides and fungicide groups. There is rather extensive information on the mechanism of resistance to the benzimidazoles, diphenylamides, carboxanilides and the phosphorothiolates, less on the DMIs and very little on the dicarboximides.

Table VII. Mechanisms of fungicide resistance[a]

Fungicide or fungicide class	Mechanisms of resistance
2-Amino-pyrimidines	Unknown
Aromatic hydrocarbons	Unknown, cross-resistance with dicarboximides
Benzimidazoles	Altered target site (β-tubulin)
Carboxanilides	Altered target site (succinate-ubiquinone oxidoreductase)
Dicarboximides	Unkown, cross-resistance with aromatic hydrocarbons
DMIs	Increased efflux; altered target site; altered demand for target-site product; target-site over-production
Dodine	Unknown
Kasugamycin	Altered target site (ribosomes)
Organo-mercurials	Detoxification by binding substances
Phenylamides	Altered target site (RNA polymerase)
Phosphorothiolates	Metabolic detoxification
Triphenyltins	Unknown

[a] Data are from reference 3.

There is clear evidence that the most common mechanism for fungicide resistance is biochemical target site alteration by mutation. Interestingly, degradation and conjugation by xenobiotic metabolism, which is extremely important in insecticide and herbicide resistance, are less prominent against fungicides. This mechanistic preference may be one of the reasons why some of the older and extensively used fungicides (copper-, organo-mercury- and sulfur-based), which are considered "multi-site inhibitors", have not developed high levels of "practical" resistance whereas most of the systemic-type fungicides, which are "single-site" or "site-specific" fungicides have evolved comparatively high levels of resistance in many fungal species. With this aspect in mind, it is also of interest that the use of fungicides that are "indirectly" fungitoxic, such as probenazole and tricyclazole, has not resulted in resistance. The influence of selective mode of action is likely the major determinant in the increased risk of resistance associated with the systemic versus the non-systemic fungicides. However, major resistance problems have appeared against a variety of non-systemic fungicides including vinclozolin, iprodione, dodine and fentin.

Occurrence and Extent

Over 30 years of practical experience with fungicide resistance problems has established that the development of resistance is largely related to the class to which the fungicide belongs. Table VIII summarizes the appearance of resistance to the most widely used fungicides or classes that has occurred during this period. By 1970, only a few instances of fungicide resistance had been determined (aromatic hydrocarbons, organo-mercurials, dodine) and then only after extensive and long-termed use. The switch to the newer systemic fungicides at this time greatly altered this pattern of resistance development. Resistance to the systemic fungicides, such as the benzimidazoles (benomyl, carbendazim, thiabendazole), the phenylamides (metalaxyl, oxadixyl) and the dicarboximides (iprodione, procymidone, vinclozolin) occurred rapidly, was broad-based and elicited wide-spread cross-resistance, both in terms of fungicides that had similar chemistry and similar modes of action (Table IX). Although most new groups of fungicides have been affected, the morpholines, fosetyl, probenazole, isoprothiolane and tricyclazole have shown only low levels of resistance. Other older materials, such as copper-based fungicides, sulfur, dithiocarbamates (mancozeb), phthalimides (captan) and chlorothalonil, have remained effective even after extensive and prolonged use. Resistance to the azoles (triadimefon, flutriafol, flusilazole) has developed much more slowly than expected and only within limited fungal species. It is too soon to speculate on the longevity of newer introductions such as the phenylpyrroles, anilinopyrimidines and strobilurins. However, the rapid development and

Table VIII. Chronological occurrence of fungicide resistance[a]

Year	Fungicide or class	Years of commercial use before resistance (approx.)	Crop disease and principal pathogens
1960	Aromatic hydrocarbons	20	Citrus storage rots, *Penicillium* spp.
1964	Organo-mercurials	40	Cereal leaf spot and stripe, *Pyrenophora* spp.
1969	Dodine	10	Apple scab, *Venturia inaequalis*
1970	Benzimidazoles	2	Many target diseases and pathogens
1971	2-Amino-pyrimidines	2	Cucumber and barley powdery mildews, *Sphaerotheca fuliginea* & *Erysiphe grisea*
1971	Kasugamycin	6	Rice blast, *Magnaporthe grisea*

1976	Phosphorothiolates	9	Rice blast, *Magnaporthe grisea*
1977	Triphenyltins	13	Sugar-beet leaf-spot, *Cercospora betae*
1980	Phenylamides	2	Potato blight and grape downy mildew, *Phytophthora* infections & *Phytophthora viticola*
1982	Dicarboximides	5	Grape grey mould, *Botrytis cinerea*
1982	DMIs	7	Cucurbit and barley powdery mildews, *Sphaerotheca fuliginea* & *Erysiphe graminis*
1885	Carboxanilides	15	Barley loose smut, *Ustilago nuda*

[a] Data are from reference 3.

Table IX. Current status of fungicide resistance[a]

Fungicide group (FRAC code)	Chemical classification	Example	Resistance status
Benzimidazole (1)	benzimidazole	benomyl	common in many fungal species, cross-resistance between group members
Dicarboximide (2)	dicarboximide	iprodine	common in *Botrytis cinerea*, some additional fungal species, cross-resistance common
DMI (3)	imidazole	imazalil	common in various fungal species, generally cross-resistant
	piperazine	triforine	
	pyridine	pyrifenox	
	pyrimidine	fenarimol	
	triazoles	bitertanol	
Phenylamide (4)	acylalanine	benalaxyl	common in Oomycete fungi, cross-resistance common
Morpholine (5)	morpholine	tridemorph	decrased powdery mildew sensitivity

Phosphorothiolate (6)	organophosphorous	iprobenfos	some resistance for specific fungi, no cross resistance
Oxathiin (7)	anilide	mepronil	some resistance for specific fungi
Hydroxypyrimidine (8)	pyrimidinol	ethirimol	powdery mildew resistance and cross-resistace
N-phenylcarbamate (10)	diethofencarb	diethofencarb	resistance known
STAR/QoI (11)	strobilurines	azoxystrobin amoxadone	resistance and cross-resistace shown
Aromatic hydrocarbons (14)	chlorophenyl	chloroneb	resistance to some fungi
Polyoxin (19)		polyoxin	resistance known
Miscellaneous (X)	Organotins	triphenyltin	resistance in *Cercospora beta*

[a] Data are from reference 3.

implementation of resistance management practices should greatly improve their usefulness as effective fungal control agents.

Management

Integrated disease management (IDM) has been defined as the combined use of all types of countermeasures against plant diseases including disease-resistant crop varieties, biological control agents, and hygienic practices such as crop rotation and removal of diseased aspects of perennial plants. As with insecticides and herbicides, the development of resistance is most evident following the sustained sole use of fungicides with the same and specific modes of action. Management practices suggested by the FRAC include; selection of fungicides with different mechanisms of action, mixture applications of different types of fungicides or as rotations, reduction in the application of at-risk fungicides, restriction in the number of applications to only those absolutely needed and change fungicide class if reapplication is necessary, use of fungicides only at their recommended dose unless the genetics of resistance are known and dictate an adjustment, avoidance of eradicant use of phenylamides for control of foliage diseases and for DMIs in some fruit crops, and increased use of non-chemical control means to minimize the need and extent of chemical intervention.

References

1. Insecticide Resistance Action Committee. **2001**. http://plantprotection.org/IRAC/general_resources/moafinal.html.
2. herbicide Resistance Action Committee. **2001**. http://plantprotection.org/HRAC/brighton.html.
3. Fungicide Resistance Action Committee. **2001**. http://www.gcpf.org/frac/chemicals%20group.html.
4. Anonymous. **1957.** 7th Rept.World Health Organ. Tech. Rept. Ser. No. 125. 32.
5. Medcalf, R.L. *Organic Insecticides;* Wiley (Interscience): New York, NY, 1955; pp 24-25.
6. O'Brien, R.D. *Insecticides*; Academic. Press: New York, NY, 1967; pp 231.
7. Ryan, G.F. *Weed Science*. **1970**, 617-616.
8. Le Baron, H.M. In *Herbicide Resistance in Weeds and Crops;* Caseley, J.C.; Cussans, G.W.; Atkin, R.K., Eds.; Oxford: Butterworth Heinemann: London, UK, 1991, pp 27-43.
9. Heaps, I. **1997.** http://www.weedscience.com.

Insecticide Resistance

Chapter 2

Monooxygenase-Mediated Insecticide Resistance: Regulation of *CYP6D1* Expression

Jeffrey G. Scott and Shinji Kasai

Department of Entomology, Comstock Hall, Cornell University, Ithaca, NY 14853

Cytochrome P450 monooxygenase-mediated detoxification is a major mechanism of resistance to insecticides. Although major advances have been made, we are still a long way from completely understanding the molecular basis of this resistance. Cytochrome P450 6D1 (CYP6D1) is responsible for monooxygenase-mediated resistance to pyrethroid insecticides in the Learn PyR (LPR) strain of house fly. The underlying mechanism for increased detoxification of pyrethroids in LPR is elevated transcription of *CYP6D1.* Identification of the DNA sequence elements (and the specific proteins that bind these sites) responsible for increased transcription of *CYP6D1* and other P450s involved in resistance will provide a vastly improved understanding of this important resistance mechanism. Some of the possible mechanisms by which an elevated rate of transcription could occur are discussed.

The cytochrome P450-dependent monooxygenases (monooxygenases) are a vital biochemical system because they metabolize xenobiotics such as drugs, pesticides and plant toxins, and because they regulate the titers of

© 2002 American Chemical Society

endogenous compounds such as hormones, fatty acids and steroids. Cytochrome P450 (P450) is a hemoprotein which acts as the terminal oxidase in monooxygenase systems and there are multiple P450s in eukaryotic species. Monooxygenases are remarkable in that they can oxidize widely diverse substrates and are capable of producing a bewildering array of reactions (1-3).

Monooxygenase-mediated oxidation requires (besides oxygen and substrate) P450, NADPH cytochrome P450 oxidoreductase (P450 reductase), NADPH and phospholipid (4). P450 reductase is required to transfer reducing equivalents from NADPH to P450. Cytochrome b_5 is involved in certain monooxygenase reactions; depending upon the P450 and/or substrate involved (5-9).

In 1987 a P450 nomenclature was proposed (10) and, although subject to revisions (11-13), it remains the preferred system (14). Sequences are named *CYP* (for cytochrome P450), followed by a number, a letter and a number indicating the family, subfamily and isoform, respectively (14). Alleles are designated *v1*, *v2*, etc. Since this nomenclature is based on overall sequence similarity, no information regarding the function of a P450 should be assumed by its classification within this system (15).

The number of P450 genes has been growing at a remarkable rate and attempts to document the current number of genes rapidly fall out of date. Sequencing of the *Caenorhabditis elegans* genome revealed 80 P450 genes (16) while 90 P450s (including 4 pseudogenes) were found in the *Drosophila melanogaster* genome (17).

One remarkable feature of the monooxygenases is the large variation in substrate specificity of different P450s. For example, CYP1A1 can metabolize more than 20 substrates, while CYP7A1 has only one known substrate (2). Certain P450s have overlapping substrate specificity (e.g. CYP2C subfamily in humans) (2) so that a single compound may be subject to metabolism by multiple P450s. In addition, some P450s produce only a single metabolite from a given substrate, while other P450s can produce multiple metabolites. These features are further complicated by the fact that the change of a single amino acid in a P450 can alter its substrate specificity (18). More information about known monooxygenase substrates can be found in several reviews (1, 2, 19-26).

Inhibition of P450s

P450 inhibitors are a diverse group of compounds that have been variously classified based on type of inhibition, type of spectrum formed, or reaction/isoform that is inhibited. P450 inhibitors can be used as insecticide synergists, with piperonyl butoxide being a well known example (27).

Induction of monooxygenases

The level of monooxygenase activity is known to increase significantly upon exposure to certain types of naturally occurring or synthetic compounds, a phenomenon known as induction. The bulk of published work on induction of monooxygenases has dealt with mammals, where inducers can be classified based on the P450 isoforms that are induced (28). Additionally, information about how a P450 is regulated can be provided by induction studies (29). It has been suggested that information about the evolutionary classification of P450s may also be deduced from induction studies, although diverse P450s can respond to similar inducers (29).

The P450 monooxygenases of insects

The detection of total P450 was first reported from an insect in 1967 (30) and subsequently many researchers found convincing evidence for multiple P450s in several insect species (refs in (19)). Although there has been one case of an orthologous gene with a high degree of similarity between insect species (31), diversity seems to be the rule for most of the >200 insect P450s thus far sequenced. Given that there may be as many as 1,000,000 species of insects, the possible number of P450s in insects becomes staggering.

Although there are multiple P450s in any given insect there is only one P450 reductase and one b_5 in each species. Insect P450 reductase was first purified and characterized from house fly (32, 33) and has been cloned from house fly (34) and *Drosophila* (35). Cytochrome b_5 was first identified in the insect, *Hyalophora cecropia* (36). Subsequently, insect b_5 was purified from house flies (37-41) and cloned from *Drosophila* (42, 43), house fly (44) and *Helicoverpa armigera* (45).

Insect monooxygenases can be detected in many tissues. Highest monooxygenase activities are usually associated with the midgut, fat bodies and Malpighian tubules (21), but expression of P450 isoforms varies (46).

Dramatic variation in the levels of monooxygenase activities and P450 levels occur during the development of most insects. In general, total P450 levels are undetectable in eggs, rise and fall in each larval instar, are undetectable in pupae and are expressed at high levels in adults (19). The patterns of expression of a P450 can vary within and/or between life stages (46-49). P450 reductase and b_5 are expressed in all life stages (19).

The monooxygenases of insects have several functional roles, including resistance to pesticides, growth, development, feeding and tolerance to plant toxins. Recent reviews cover some of these topics in detail (46, 50-52). Identification of the P450s involved in these important processes is currently an area of intensive research.

Monooxygenase-Mediated Insecticide Resistance

The most common types of resistance found in insects are increased enzymatic detoxification and target site insensitivity (19, 53, 54). It has been postulated (22, 53, 54) that increased enzymatic detoxification is the most frequently occurring resistance mechanism, but altered target sites (55) are also very common. Monooxygenase-mediated resistance may be the most common type of metabolism based insecticide resistance (53, 54, 56, 57), although esterases are also important (58), and glutathione S-transferases are significant in some cases (59). In addition, monooxygenase-mediated detoxification in susceptible strains of certain species substantially limits the toxicity and usefulness of some insecticides, such as pyrethrins (60), imidacloprid (61) and carbaryl (62). Furthermore, P450s are responsible for the activation of many organophosphate insecticides (63).

Most cases of monooxygenase-mediated resistance result from an increase in detoxification. However, in cases where the parent insecticide must undergo monooxygenase-mediated bioactivation, as is the case for many organophosphates, it is also possible that resistance could be achieved through decreased activation. Although this has been reported once (64), it does not appear to be a common mechanism of resistance. This may explain why esterases are relatively more common than monooxygenases in resistance to organophosphates in several species (53, 54).

The realization that there were multiple P450s in insects led to the suggestion that monooxygenase-mediated resistance could result from an increased expression of the P450(s) responsible for metabolism of the insecticide, or from a change in the protein (resulting in a higher turn over of the insecticide (53)). As our understanding of gene regulation has increased over the last 15 years so has the list of possible mechanisms for accomplishing an increase in expression (or a change in the catalytic activity). For example, increased expression could occur (at least theoretically) by increased transcription, gene amplification, stabilization of the mRNA, stabilization of the protein, etc. In fact, there are numerous ways that some of these mechanisms (e.g. increased transcription) could be accomplished.

It is now clear that resistance can occur by increased transcription of a P450 leading to both increased expression of the protein and increased detoxification of the insecticide (65). Interestingly, this can occur by *cis* and *trans* regulation of the transcription process (65). It is also noteworthy that the factor(s) responsible for *trans* regulation of the P450(s) involved in resistance also appear capable of regulating the expression of other P450s that are not necessarily involved in resistance (at least in house flies). This adds another level of complexity to the study of insect P450s, because not only are the substrate specificities variable, but regulation of P450 genes are also subject to controls that will vary. Knowing which subset of P450s are elevated by the same regulatory factor(s) will help us understand cross-resistance patterns.

One consistent result regarding the constitutive overexpression of P450s in resistant strains (independent of their role of the P450 in resistance) is that gene amplification does not appear to be involved (52). This is in contrast to resistance mediated by esterases in which gene amplification appears to be the major mechanism for overexpression of these proteins (58, 66, 67). The lack of P450 gene amplification in insects is even more surprising given that there are known cases of P450 gene amplification that result in individuals with increased metabolism of specific compounds in humans (68).

Increased levels of P450 reductase (69) and b_5 (70) were first associated with monooxygenase-mediated insecticide resistance in house flies, and this observation has been confirmed in other species (71-74). The discovery of elevated levels of P450 reductase and b_5 associated with monooxygenase-mediated resistance led to the suggestion that elevated levels of one or both of these proteins could be involved in resistance (70).

It is now known that resistance can occur due to detoxification via a single P450 (75) and that the metabolic attack can be limited to a single site on the insecticide (76). It is also clear that b_5 is needed for P450-mediated detoxification of some pesticides (76, 77), and is involved in at least one case of insecticide resistance (6).

One point of clarification which must be emphasized is that many researchers focus on P450s that are overexpressed in a resistant strain as the only ones that could be involved in resistance. At this point in time, such an exclusion of other P450s seems inappropriate, because of the possibility that a change in the P450 amino acid sequence could also result in insecticide resistance. By analogy, a recent report demonstrated that organophosphate resistance in blowfly could be conferred by a change in the gene coding for a carboxylesterase (78). In the susceptible strain this enzyme metabolizes α-naphthylacetate, but not organophosphates. The resistant strain has a mutant form of this enzyme which no longer metabolizes α-naphthylacetate, but effectively metabolizes organophosphates (78). It is possible that such a mechanism could occur for monooxygenase-mediated insecticide resistance, especially since it is known that a single amino acid change can substantially alter the substrate specificity of a P450 (18). Therefore, to decide a P450 is not involved in resistance because it is not expressed at higher levels in a resistant strain may not be accurate.

CYP6D1-mediated pyrethroid resistance in LPR house flies

Research in our laboratory has focused on the molecular basis of monooxygenase-mediated resistance to pyrethroid insecticides in the house fly. We have identified the P450 involved in this resistance (CYP6D1) and what is known about CYP6D1-mediated resistance will be elaborated on below. A recent review covers information on what is known about the role of other P450s in insecticide resistance (52).

The LPR strain of house fly was collected in 1982 from a dairy in New York following the introduction of permethrin for fly control. After permethrin selection the LPR strain became homozygous for the major mechanisms of resistance and attained extremely high levels of resistance to commonly used pyrethroid insecticides (70). The highest levels of resistance occur toward pyrethroids with a phenoxybenzyl moiety (e.g. >5,000-fold resistance to cypermethrin)(70, 79).

Monooxygenase-mediated pyrethroid resistance in LPR has been demonstrated by *in vivo* and *in vitro* studies (70, 75, 80-82). The two other mechanisms of resistance to pyrethroid insecticides in the LPR strain are *kdr* and decreased cuticular penetration (*pen*) (70, 82, 83). Monooxygenase-mediated detoxification appears to be the major mechanism of pyrethroid resistance in this strain (70).

Detailed genetic analysis of permethrin resistance revealed that the high level of piperonyl butoxide suppressible resistance (i.e. P450-mediated detoxification) was associated with autosomes 1 and 2 in combination, neither autosome having a substantial effect by itself (82). Both *kdr* and *pen* were found on autosome 3 (82, 83).

A single P450, CYP6D1 (a.k.a. $P450_{lpr}$), was purified from LPR house flies by HPLC (84). Electrophoresis of house fly microsomes indicated that a protein band corresponding to CYP6D1 was expressed at elevated levels in LPR compared to susceptible house flies (85). An isoform specific antiserum was raised in rabbits using purified CYP6D1 protein as the antigen (86) and this antiserum was used to characterize the expression of CYP6D1 (see below). The *CYP6D1* gene has been sequenced (87) and cloned (88).

To ascertain the role of CYP6D1 in pyrethroid resistance in the LPR strain, the P450-dependent metabolism of deltamethrin was investigated. Monooxygenase dependent deltamethrin metabolism occurred at increased levels in LPR microsomes compared to a susceptible strain, and this enhanced metabolism of deltamethrin was eliminated by the addition of anti-CYP6D1 antiserum. Thus, CYP6D1 is the major P450 responsible for deltamethrin metabolism in LPR flies (75). Similar results were observed for cypermethrin, and the primary CYP6D1 specific metabolite formed *in vitro* was identified as 4'-OH cypermethrin (76). Thus, CYP6D1 carries out metabolism at a single site on the pyrethroid phenoxybenzyl moiety. This helps to explain the reduced levels of resistance in LPR to pyrethroids lacking this functional group (70).

Immunoinhibition studies demonstrated that b_5 is required for CYP6D1-mediated metabolism of cypermethrin in microsomes from LPR house flies (6). Therefore, b_5 appears to be directly involved in CYP6D1-mediated pyrethroid resistance. The higher level of b_5 found in the LPR strain (83) and the linkage of this trait to the same autosomes as monooxygenase-mediated resistance (81) suggests that the elevated levels of b_5 may be a requisite for the enhanced metabolism of pyrethroids by CYP6D1 in LPR house flies.

Northern blots of RNA from adults of the LPR and an insecticide susceptible strain revealed that CYP6D1 mRNA was expressed at about a 10-fold higher level in LPR flies compared to susceptible flies. This agreed with previous results of an 8-fold higher level of CYP6D1 protein in microsomes from LPR relative to susceptible flies (81, 86).

Southern blots of genomic DNA hybridized with a *CYP6D1* cDNA revealed similar hybridization intensities between LPR and a susceptible strain, indicating that the elevated level of CYP6D1 mRNA in LPR flies is not due to gene amplification (87). *CYP6D1* was mapped to chromosome 1 (89).

To investigate the role of mRNA stability in the high level expression of CYP6D1 in the LPR strain, mRNA was isolated and quantified from aabys (susceptible) and LPR house flies at different times following injection of a transcription inhibitor (actinomycin D). The same pattern of decrease in CYP6D1 mRNA abundance (approximate half-life of ~10 hr) was detected in both LPR and aabys strains, indicating that the increased expression of *CYP6D1* in LPR is not due to increased stability of the mRNA (65).

The relative transcription rate of *CYP6D1* in LPR and a susceptible strain was measured using an *in vitro* run-on transcription assay. When nuclei from the susceptible strain were used in the run-on assay only a trace of CYP6D1 mRNA was detected (65). In contrast, when nuclei from the LPR strain were used for the run-on assay abundant CYP6D1 mRNA was detected. Quantitation of the relative intensities of the CYP6D1 signals between aabys and LPR revealed approximately a 10-fold difference (65) which is comparable to the differences in *CYP6D1* expression observed by Northern hybridization (81, 90). This was the first direct evidence for increased transcription as an underlying cause of insecticide resistance (65).

The increased rate of transcription of *CYP6D1* was due to factors on autosomes 1 and 2 in the LPR strain (i.e. both by *cis* and *trans* regulatory factors), with the factor on autosome 2 having the greater effect (65). This is consistent with what is known about the linkage of monooxygenase-mediated resistance (82, 83) and overexpression of CYP6D1 mRNA and protein levels (81). Curiously, the level of monooxygenase mediated resistance associated with autosome 1 was found to be slightly greater than that associated with autosome 2 (82), suggesting that resistance associated with autosome 1 is due to increased transcription and some other factor.

In addition to the LPR strain, *CYP6D1* cDNA was sequenced from four pyrethroid susceptible strains of house flies. Comparison of the five *CYP6D1* alleles reveals that the deduced amino acid sequence from the LPR allele differs from that of the CS, aabys, ISK and Rutgers (strain not homozygous) alleles by 8, 11, 7 and 6-7 amino acids, respectively (90). Among them, 5 amino acids are the same in CS, aabys, ISK and Rutgers, but are different from LPR: Asp_{150} to Ala, Ile_{153} to Leu, Thr_{165} to Ser, Glu_{218} to Gln and Met_{227} to Ile (90). The observed amino acid substitutions occur at two highly variable regions among cytochromes P450 in family 6, and the changes at residues 218 and 227 are close to a proposed substrate binding region (91). In addition, we recently sequenced a partial clone of *CYP6D1* from a genomic library prepared from the susceptible Edinburgh strain of house fly (obtained from Mary Bownes, U.K.). The first 229 amino acids have 7 differences from LPR, including the same amino acid substitutions found for the other four pyrethroid susceptible strains (unpublished results). Thus, pyrethroid susceptible strains of house fly from Japan, Europe and North America have *CYP6D1* gene sequences that are more closely related to each other, than any are to LPR. This suggests there is a selective advantage for LPR in having this particular allele.

Understanding which compounds can be metabolized by CYP6D1 offers insight into the structure of the active site of this P450 and provides information about possible cross-resistance patterns. CYP6D1 carries out metabolism of methoxyresorufin, benzo[*a*]pyrene (92), chlorpyrifos (93), phenanthrene (94) and pyrethroids (6, 75, 95), but it does not substantially metabolize ethoxyresorufin, pentoxyresorufin or ethoxycoumarin (92). Thus, the active site of CYP6D1 will readily accommodate substrates with considerably different structures, yet at the same time small changes in the structure of a substrate can radically reduce the rate at which the compound is metabolized. A similar trend in substrate specificity has been observed for CYP6A1 (96, 97). In the case of CYP6D1 some substrates, such as methoxyresorufin, b_5 is not required. Yet for other substrates, such as benzo[*a*]pyrene and pyrethroids, b_5 is required (6, 76).

General P450 inhibitors have found utility as insecticide synergists. To evaluate whether or not we could identify specific inhibitors of CYP6D1, and to determine their ability to act as insecticide synergists, >30 compounds have been examined (98, 99). Several potent inhibitors of CYP6D1 were discovered, including xanthotoxin, chlorpyrifos, 4-ethynylpyrene and 9,10-methylenedioxy phenanthrene. Some compounds, such as isosafrole, verbutin and 9,10-methylenedioxy phenanthrene, were also shown to be potent synergists (98, 99). Generally the more potent inhibitors were large planar compounds (99). Results from these studies clearly indicate that identification of isoform selective inhibitors of P450s within an insect, and across species, is possible. This suggests that it should be possible to design pesticide synergists that can effectively increase the toxicity to pest species without a concomitant increase in toxicity to non-target organisms.

Factors Regulating CYP6D1 Expression

CYP6D1 mRNA expression is developmentally regulated with no CYP6D1 mRNA detectable in eggs, larvae, or pupae (49). High levels of mRNA were found in adults from 1 to 6 days old. This pattern matches that observed for the CYP6D1 protein (100). Results from studies on individual tissues indicate that CYP6D1 protein is found throughout the house fly abdomen (101, 102). CYP6D1 is also expressed in all tagmata and in thoracic ganglia of house fly (95). The level of expression of CYP6D1 at each of these tissues and/or sites was higher in LPR compared to susceptible strains (101, 102). This suggests there is no single tissue or site within the LPR house fly that is responsible for pyrethroid resistance. Furthermore, monooxygenase-mediated detoxification at the level of the target tissue may help to explain the high levels of resistance to pyrethroids in the LPR strain.

To better understand the regulation of *CYP6D1*, the effects of 21 monooxygenase inducers were examined (49). In adult flies, the highest level of induction was seen for phenobarbital. Total P450s were inducible by phenobarbital in CS house fly larvae, but not in LPR larvae. Northern blots of phenobarbital treated CS flies indicated that there was a 4-fold increase in CYP6D1 mRNA levels over the untreated flies. In the LPR strain there was no induction of CYP6D1 by phenobarbital (49, 101, 102). Following phenobarbital induction the level of CYP6D1 mRNA in the CS strain was about half of the level in the LPR strain. Immunoblotting revealed no detectable CYP6D1 in control or PB-treated larvae in either strain (49). Thus, although total P450s are inducible in the susceptible strain larvae with PB, CYP6D1 is not.

In summary, CYP6D1 expression is 1) found in all tissues of adult flies at elevated levels in LPR relative to susceptible strains, 2) elevated in LPR is due to factors on autosomes 1 and 2, 3) expressed only in adults and 3) is inducible by phenobarbital in susceptible house fly adults, but not in LPR.

Transcriptional Regulation of CYP6D1

A major unanswered question about monooxygenase-mediated resistance is what controls the transcription of the P450(s) involved. In this section, we discuss the possible mechanisms of *CYP6D1* gene regulation based on what is known about the regulation of other P450 genes.

DNA sequence elements and their interaction with specific proteins

The most common type of gene regulation occurs due to the presence or absence of 1) regulatory elements and/ or 2) the proteins that bind to these elements. These elements are most commonly found in the 5′ flanking sequence of the genes they regulate, although they can be found elsewhere as well (i.e. 3' flanking sequence, introns, etc.)(103). In mammals, several different transcription factors have been identified (HNF-1α, HNF-3, HNF-4, C/EBPβ,

Sp1, GABP α/β, NF2d9, etc.) as having a role in regulation of P450 gene expression (104). To gain information about possible regulatory elements involved in *CYP6D1* expression a house fly genomic library was screened. A *CYP6D1v1* clone was isolated and sequenced. This clone contained 887 nucleotides 5′ to the open reading frame. Using PCR with primers based on the *CYP6D1v1* allele, the sequences 5′ to the ORF were obtained from five pyrethroid susceptible strains. The transcription initiation site was identified at the same position in LPR and susceptible strains (86 nucleotides upstream from the translation start site). Comparison of the 5′ flanking sequences revealed a high degree of similarity for most regions, although differences in the sequences were identified (105). None of the known P450 regulatory sequences were found within the *CYP6D1* 5′ flanking sequence (105). The most notable difference in the 5′ flanking sequence of *CYP6D1* between strains was the presence of 15-bp fragment close to the transcriptional initiation site which was found in the LPR strain, but was absent in all of the susceptible strains. A search of GenBank and the TRANSFAC database revealed that the sequence in this region from susceptible strains (without the 15-bp fragment) shows high similarity to the *Gfi-1* DNA binding domain. Binding of *Gfi*-1 to this consensus sequence inhibits transcription in rats (106). Thus, it is possible that the 15-bp fragment in the LPR strain acts to disrupt a *Gfi*-1-like repressor from binding; leading to constitutively high levels of *CYP6D1* expression in this strain. However, no *Gfi*-1 homologous protein has yet been identified from insects, so the role of this 15-bp fragment in regulation of *CYP6D1* will require further study.

Several regulatory elements in the 5′ flanking region of the human aromatase (*CYP19*) gene have been identified (107). One of the important elements, a guanine and cytosine enriched sequence (GC-box), was identified at about –230 bp from the first exon (107). A nuclear protein, called Sp1, binds this site and increases *CYP19* gene expression (107). *CYP6D1* also has GC-rich regions within 100 bp from the transcriptional initiation site and some of the nucleotides are different only in the LPR strain, as compared to five pyrethroid susceptible strains (105). Thus, it is possible that a Sp1 homologous protein binds to this region and regulates *CYP6D1* expression in housefly, analogous to the case for the human *CYP19* gene (107).

There is relatively little known about the regulatory elements of insect cytochrome P450 genes. Basal and xanthotoxin inducible transcription of *CYP6B3* from *Papilio polyxenes* has been mapped to nucleotides –1 to –838 upstream of this gene using CAT fusion constructs (108). Possible xenobiotic-responsive elements have been identified upstream of *CYP6B4v2*, *CYP6B3v2* and *CYP6B5v1* (109) and possible antioxidant-responsive elements have been found upstream of *CYP6B1v3*, *CYP6B4v2* and *CYP6B5v1* (109). No homologous sequences for any of these regulatory elements have been found for *CYP6D1*.

DNA methylation, histone acetylation and chromatin structure

DNA methylation is an important mechanism in the regulation of gene expression (110). In animal DNA, from 2-7% of the cytosines are speculated to be methylated (103). In higher order eukaryotes, only 5′ cytosines of CpG dinucleotide (CCGG) can be methylated (111). Methylation of only one or two of the CpG sites in the promoter region causes inactivation of a gene due to inhibition of nuclear protein binding (112). For example, binding to the xenobiotic responsive element (XRE) of the AhR and Arnt complexes was inhibited by the hypermethylation of CpG dinucleotides within an XRE core sequence, and the gene transcription of rabbit *CYP1A1* was repressed (113).

DNA methylation is also associated with tissue, sex and life stage specific expression of cytochrome P450s. *CYP2E1* is expressed at a very low level in the human lung and kidney while variable level of expression in full-term placentas. In the lung and kidney, genomic DNA is heavily methylated at the CpG site of *CYP2E1* promoter region and probably it causes inhibition of binding of nuclear factors (114). Methylation is associated with both developmental and sex specific expression of mouse Cyp2d9 (115). Furthermore, the initial transcription of the hepatic P450s, *CYP2D3* and *CYP2D5* are different but are all regulated by the DNA methylation (116). Although two possible methylation sites (CCGG) are found within 100 bp from transcriptional initiation site in the upstream sequence of *CYP6D1* gene (105), it is unclear if DNA methylation is associated with the gene regulation of *CYP6D1*.

Chromatin is composed primarily of DNA and structural proteins such as histones. In order for transcription to occur, the chromatin must become modified so as to allow access of various transcription factors to the DNA. Each core histone has two domains: 1) a histone fold domain that is involved in interactions with other histones and in wrapping of DNA and 2) and an amino-terminal tail domain that lies on the outside of the nucleosome where it can interact with DNA and regulatory proteins. The amino terminal tail domains are targets for acetylation which reduces the affinity for DNA. Thus, high levels of histone acetylation correlate with gene activity and reduced levels are associated with gene silencing. Covalent modification of the components of chromatin, by DNA methylation and/or histone acetylation, provides an feasible means of regulating gene activity and may be one of the means by which CYP6D1 expression is regulated.

Phenobarbital induction

Phenobarbital is a well studied inducer of P450s. Cytochrome P450 BM1 and BM3 are phenobarbital inducible P450s from *Bacillus megaterium.*

The nuclear protein BM3R1, whose coding region is located just upstream of the P450 *BM3* gene, binds to the promoter region of P450 BM3 called the "Barbie box" and depresses transcription of P450 BM3 (117). The 5′ flanking sequence of several phenobarbital inducible P450s contain Barbie box-like sequences (118). Whether or not Barbie boxes are involved in phenobarbital induction in bacteria remains controversial (119) (120).

In insects, possible Barbie box sequences have been identified upstream from *CYP6A1*, *CYP6A2* (121), *CYP6B1v3*, *CYP6B4v2* and *CYP6B5v1* (109). As a preliminary step towards identifying whether Barbie boxes could be involved in phenobarbital induction of *CYP6D1*, sequences that were similar to the Barbie box consensus (ATCAAAAGCTGGAGG, (118)) were identified from the susceptible strain alleles and then compared with the non-phenobarbital inducible LPR strain. Two sequences matching 7-8 of the 15 nucleotides which comprise the Barbie box consensus sequence (118) were identified (105). For both putative Barbie box sequences the LPR strain differed from the phenobarbital responsive strains by 1-2 nucleotides.

It is uncertain how relevant the bacterial phenobarbital induction system is to those of mammals or insects. Rat *CYP2B1* and *CYP2B2* genes have Barbie box-like sequences in their upstream region which bind barbiturate dependent nuclear protein (119). In contrast to bacteria, the binding of the rat proteins was strongly and specifically increased with protein extracts from phenobarbital-treated rats suggesting that the Barbie box responsive element acts as a positive element (122). Furthermore, phenobarbital induction in mammals appears to be controlled by factors other than Barbie box responsive elements (123-125).

CYP6D1 is induced by phenobarbital treatment in susceptible housefly strains, but not in LPR (49, 126). It is also important to note that both phenobarbital responsiveness and CYP6D1-mediated insecticide resistance map to autosome 2 in house fly. Given the original suggestion of Terriere that "the same regulatory genes may be involved in both induction and biochemical resistance" (127, 128) and results with CYP6D1 that are consistent with this idea, resolution of the mechanism(s) of phenobarbital induction of *CYP6D1* may lead us closer to the resolution of the mechanism of *CYP6D1* over-expression in LPR strain.

Unanswered Questions

In addition to identification of the DNA sequence elements (and the specific proteins that bind these sites) responsible for increased transcription of the P450s involved in resistance, other important questions remain. For example, do different P450 alleles have a role in resistance? Another unresolved question is why mRNA levels can be considerably increased while the level of total P450s increases only slightly? For example, in LPR total P450 protein levels are only elevated by 2-4 fold. Yet mRNAs for CYP6D1 (49, 81, 90), CYP6A1(129) and CYP6D3 (unpublished data) are each increased by 8- to 15-fold each. Thus, while CYP6D1 appears to be efficiently transcribed into P450 holoenzyme (CYP6D1 protein and mRNA levels are both about 10-fold elevated in LPR), either CYP6D3 and/or CYP6A1 are not, or there is substantial coordinate down regulation of other P450s such that the level of total P450s does not increase in proportion with the increase in mRNA. Studies aimed at examining these questions will greatly enhance our understanding of insect P450s and monooxygenase-mediated resistance.

Acknowledgements

This work was supported by a grant from the National Institutes of Health (GM 47831).

References

1. A. P. Kulkarni, E. Hodgson; *Pharmac. Ther.* **1980,** 8, 379-475.
2. S. Rendic, F. J. Di Carlo; *Drug Metab. Rev.* **1997,** 29, 413-580.
3. D. Mansuy; *Comp. Biochem. Physiol.* **1998,** 121C, 5-14.
4. A. Y. H. Lu, M. J. Coon; *J. Biol. Chem.* **1968,** 243, 1331-1332.
5. J. A. Peterson, R. A. Prough, In *Cytochrome P-450 Structure, Mechanism, and Biochemistry*; Ortiz de Montellano, P. R., Eds.; Plenum Press: New York and London, 1986; pp 89-117.
6. M. Zhang, J. G. Scott; *Pestic. Biochem. Physiol.* **1996,** 55, 150-156.
7. K. P. Vatsis; D. P. Gurka, P. F. Hollenberg, In *Biochemistry, Biophysics, and Regulation of Cytochrome P-450*; Gustafsson, J., Carlstedt-Duke, J., Mode, A. and Rafter, J., Eds.; Elsevier: New York, 1980; pp 347-350.
8. D. Pompon; *Biochem.* **1987,** 26, 6429-6435.
9. P. M. Epstein; M. Curti; I. Jansson; C.-K. Huang, J. B. Schenkman; *Arch. Biochem. Biophys.* **1989,** 271, 424-432.
10. D. W. Nebert; M. Adesnik; M. J. Coon; R. W. Estabrook; F. J. Gonzalez; P. F. Guengerich; I. C. Gunsalus; E. F. Johnson; B. Kemper; W. Levin; I. R. Phillips; R. Sato, M. R. Waterman; *DNA* **1987,** 6, 1-11.
11. D. W. Nebert; D. R. Nelson; M. Adesnik; M. J. Coon; R. W. Estabrook; F. J. Gonzalez; P. F. Guengerich; I. C. Gunsalus; E. F. Johnson; B. Kemper; W. Levin; I. R. Phillips; R. Sato, M. R. Waterman; *DNA* **1989,** 8, 1-13.
12. D. W. Nebert; D. R. Nelson; M. J. Coon; R. W. Estabrook; R. Feyereisen; Y. Fujii-Kuriyama; F. J. Gonzalez; F. P. Guengerich; G. I. C.; E. F.

Johnson; J. C. Loper; R. Sato; M. R. Waterman, D. J. Waxman; *DNA Cell Biol.* **1991,** 10, 1-14.

13. D. R. Nelson; T. Kamataki; D. J. Waxman; F. P. Guengerich; R. W. Estabrook; R. Feyereisen; F. J. Gonzalez; M. J. Coon; I. C. Gunsalus; O. Gotoh; K. Okuda, D. W. Nebert; *DNA Cell Biol.* **1993,** 12, 1-51.
14. D. R. Nelson; L. Koymans; T. Kamataki; J. J. Stegeman; R. Feyereisen; D. J. Waxman; M. R. Waterman; O. Gotoh; M. J. Coon; R. W. Estabrook; I. C. Gunsalus, D. W. Nebert; *Pharmacogenetics* **1996,** 6, 1-42.
15. D. R. Nelson; *http://drnelson.utmem.edu/nelsonhomepage.html* **1998**.
16. *C. elegans. Consortium*; *Science* **1998,** 282, 2012-2018.
17. M. D. Adams; S. E. Celniker; R. A. Holt; C. A. Evans; J. D. Gocayne, *et al.*; *Science* **2000,** 287, 2185-2195.
18. R. L. P. Lindberg, M. Negishi; *Nature* **1989,** 339, 632-634.
19. M. Agosin, In *Comprehensive Insect Physiology, Biochemistry, and Pharmacology*; Kerkut, G. A. and Gilbert, L. I., Eds.; Pergamon Press: New York, 1985; Vol. 12, pp 647-712.
20. E. Hodgson, In *Comprehensive Insect Physiology Biochemistry and Pharmacology*; Kerkut, G. A. and Gilbert, L. C., Eds.; Pergamon Press: Oxford, 1985; Vol. 11, pp 647-712.
21. E. Hodgson; *Insect Biochem.* **1983,** 13, 237-246.
22. C. F. Wilkinson, In *Pest Resistance to Pesticides*; Georghiou, G. P. and Saito, T., Eds.; Plenum Press: New York, 1983; pp 175-206.
23. C. F. Wilkinson, In *Bioregulators for Pest Control*; Hedin, P. A., Eds.; American Chemical Society: Washington D.C., 1985; Vol. ACS Symp. Ser. 276, pp 161-176.
24. S. Ahmad; *J. Chem. Ecol.* **1986,** 12, 533-539.
25. F. P. Guengerich; *Chemico-Biol. Interact.* **1997,** 106, 161-182.
26. F. P. Guengerich, In *Cytochrome P450 Structure, Mechanism, and Biochemistry*; Ortiz de Montellano, P. R., Eds.; Plenum Press: San Francisco, 1995; pp 473-535.
27. *Piperonyl Butoxide: The Insecticide Synergist*; D. G. Jones; Academic Press: San Diego, 1998; pp 1-323.
28. R. Snyder, R. H., In *Hepatic Cytochrome P-450 Monooxygenase System*; Schenkman, J. B. and Kupfer, D., Eds.; Pergamon: New York, 1982; pp 227-268.
29. D. W. Nebert; D. R. Nelson, R. Feyereisen; *Xenobiotica* **1989,** 19, 1149-1160.
30. J. W. Ray; *Biochem. Pharmacol.* **1967,** 16, 99-107.
31. L. Xianchun; M. R. Berenbaum, M. A. Schuler; *Insect Biochem. Molec. Biol.* **2000,** 30, 75-84.
32. T. G. Wilson, E. Hodgson; *Insect Biochem.* **1971,** 1, 171-180.
33. T. G. Wilson, E. Hodgson; *Insect Biochem.* **1971,** 1, 19-26.
34. J. F. Koener; F. A. Carino, R. Feyereisen; *Insect Biochem. Molec. Biol.* **1993,** 23, 439-447.

35. B. T. Hovemann; F. Sehlmeyer, J. Malz; *Gene* **1997,** 189, 213-219.
36. R. C. Sanborn, C. M. Williams; *J. Gen. Physiol.* **1950,** 33, 579-588.
37. Y. Okada, K. Okunuki; *J. Biochem.* **1970,** 67, 487-496.
38. Y. Okada, K. Okunuki; *J. Biochem.* **1969,** 65, 581-596.
39. G. D. Wheelock, J. G. Scott; *Comp. Biochem. Physiol.* **1992,** 101B, 209-215.
40. A. Megias; A. Saborido, A. M. Municio; *Bichim. Biophy. Acta* **1986,** 872, 116-124.
41. K. A. McFadden; D. L. Crankshaw; H. K. Hetnarski, C. F. Wilkinson; *Insect Biochem.* **1979,** 9, 301-308.
42. R. J. Levin; P. L. Boychuk; C. M. Croniger; J. A. Kazzaz, C. E. Rozek; *Nucl. Acids Res.* **1989,** 17, 6349-6367.
43. M. E. Kula; E. R. Allay, C. E. Rozek; *J. Molec. Evol.* **1995,** 41, 430-439.
44. V. M. Guzov; H. L. Houston; M. B. Murataliev; F. A. Walker, R. Feyereisen; *J. Biol. Chem.* **1996,** 271, 1-9.
45. C. Ranasinghe, A. A. Hobbs; *Insect Biochem. Molec. Biol.* **1999,** 29, 145-151.
46. J. G. Scott; N. Liu, Z. Wen; *Comp. Biochem. Physiol.* **1998,** 121C, 147-155.
47. M. B. Cohen, R. Feyereisen; *DNA Cell Biol.* **1995,** 14, 73-82.
48. M. J. Snyder; J. L. Stevens; J. F. Andersen, R. Feyereisen; *Arch. Biochem. Biophys.* **1995,** 321, 13-20.
49. J. G. Scott; P. Sridhar, N. Liu; *Arch. Insect Biochem. Physiol.* **1996,** 31, 313-323.
50. R. Feyereisen; *Ann. Rev. Entomol.* **1999,** 44, 507-533.
51. M. Schuler; *Plant Physiol.* **1996,** 112, 1411-1419.
52. J. G. Scott; *Insect Biochem. Molec. Biol.* **1999,** 29, 757-777.
53. F. J. Oppenoorth, In *Comprehensive Insect Physiology, Biochemistry, and Pharmacology*; Kerkut, G. A. and Gilbert, L. I., Eds.; Pergamon Press: Oxford, 1985; Vol. 12, pp 731-774.
54. J. G. Scott, In *Handbook of Pest Management in Agriculture*; Pimentel, D., Eds.; CRC Press: Boca Raton, 1991; Vol. 2, pp 663.
55. R. H. ffrench-Constant; *Insect Biochem. Molec. Biol.* **1999,** 29, 397-403.
56. E. Hodgson, A. P. Kulkarni, In *Pest Resistance to Pesticides*; Georghiou, G. P. and Saito, T., Eds.; Plenum Press: New York, 1983; pp 207-228.
57. L. B. Brattsten; C. W. Holyoke, Jr.; J. R. Leeper, K. F. Raffa; *Science* **1986,** 231, 1255-1260.
58. J. Hemingway, S. H. P. P. Karunarantne; *Med. Vet. Entomol.* **1998,** 12, 1-12.
59. S. J. Yu; *Zool. Stud.* **1996,** 35, 9-19.
60. R. M. Sawicki; *J. Sci. Food Agric.* **1962,** 13, 283-291.
61. Z. Wen, J. G. Scott; *Pestic. Sci.* **1997,** 49, 367-371.
62. C. F. Wilkinson; *J. Agric. Food Chem.* **1967,** 15, 139-147.

63. E. Hodgson; I. S. Silver; L. E. Butler; M. P. Lawton, P. E. Levi, In *Handbook of Pesticide Toxicology*; Hayes, W. J., Jr. and Laws, E. R., Jr., Eds.; Academic Press Inc.: New York, 1991; Vol. 1, pp 107-168.
64. T. Konno; E. Hodgson, W. C. Dauterman; *Pestic. Biochem. Physiol.* **1989,** 33, 189-199.
65. N. Liu, J. G. Scott; *Insect Biochem. Molec. Biol.* **1998,** 28, 531-535.
66. C. Mouches; N. Pasteur; J. B. Berge; O. Hyrien; M. Raymond; B. R. De Saint Vincent; M. De Silvestri, G. P. Georghiou; *Science* **1986,** 233, 778-780.
67. L. M. Field, A. L. Devonshire, In *Resistance 91*; Denholm, I., Devonshire, A. and Holloman, D. W., Eds.; Elsevier: New York, 1991; pp 240-250.
68. M. Ingelman-Sundberg; *Drug Metab. Rev.* **1999,** 31, 449-459.
69. D. R. Vincent; A. F. Moldenke; D. E. Farnsworth, L. C. Terriere; *Pestic. Biochem. Physiol.* **1985,** 23, 171-181.
70. J. G. Scott, G. P. Georghiou; *Pestic. Sci.* **1986,** 17, 195-206.
71. A. C. Kotze, B. E. Wallbank; *J. Econ. Entomol.* **1996,** 89, 571-576.
72. A. C. Kotze; *Pestic. Biochem. Physiol.* **1993,** 46, 65-72.
73. S. M. Valles, S. J. Yu; *J. Econ. Entomol.* **1996,** 89, 21-26.
74. C. N. Sun; Y. C. Tsai, F. M. Chiang, In *Molecular Mechanisms of Insecticide Resistance: Diversity Among Insects*; Mullin, C. A. and Scott, J. G., Eds.; ACS: Washington D.C., 1992; Vol. 505 (ACS Symposium Series), pp 149-167.
75. G. D. Wheelock, J. G. Scott; *Pestic. Biochem. Physiol.* **1992,** 43, 67-77.
76. M. Zhang, J. G. Scott; *Arch. Insect Biochem. Physiol.* **1994,** 27, 205-216.
77. B. C. Dunkov; V. M. Guzov; G. Mocelin; F. Shotkoski; A. Brun; M. Amichot; R. H. ffrench-Constant, R. Feyereisen; *DNA Cell Biol.* **1997,** 16, 1345-1356.
78. R. D. Newcomb; P. M. Campbell; D. L. Ollis; E. Cheah; R. J. Russell, J. G. Oakeshott; *Proc. Natl. Acad. Sci. USA* **1997,** 94, 7464-7468.
79. J. G. Scott, G. P. Georghiou; *J. Econ. Entomol.* **1985,** 78, 316-319.
80. N. Liu, J. G. Scott; *J. Econ. Entomol.* **1997,** 90, 1478-1481.
81. N. Liu, J. G. Scott; *Biochem. Genet.* **1996,** 34, 133-148.
82. N. Liu, J. G. Scott; *Pestic. Biochem. Physiol.* **1995,** 52, 116-124.
83. J. G. Scott, G. P. Georghiou; *Biochem. Genet.* **1986,** 24, 25-37.
84. G. D. Wheelock, J. G. Scott; *Insect Biochem.* **1989,** 19, 481-488.
85. J. G. Scott, S. S. T. Lee; *Arch. Insect Biochem. Physiol.* **1993,** 24, 1-19.
86. G. D. Wheelock, J. G. Scott; *Pestic. Biochem. Physiol.* **1990,** 38, 130-139.
87. T. Tomita, J. G. Scott; *Insect Biochem. Molec. Biol.* **1995,** 25, 275-283.
88. F. F. Smith, J. G. Scott; *Insect Biochem. Molec. Biol.* **1997,** 27, 999-1006.
89. N. Liu; T. Tomita, J. G. Scott; *Experientia* **1995,** 51, 164-167.
90. T. Tomita; N. Liu; F. F. Smith; P. Sridhar, J. G. Scott; *Insect Molec. Biol.* **1995,** 4, 135-140.

91. O. Gotoh, Y. Fujii-Kuriyama, In *Frontiers in Biotransformation*; Ruckpaul, K. and Rein, H., Eds.; Taylor and Francis: New York, 1989; Vol. 1, pp 195-243.
92. G. D. Wheelock, J. G. Scott; *J. Exp. Zool.* **1992,** 264, 153-158.
93. R. Hatano, J. G. Scott; *Pestic. Biochem. Physiol.* **1993,** 45, 228-233.
94. P. J. Korytko; F. W. Quimby, J. G. Scott; *J. Biochem. Molec. Toxicol.* **2000,** 14, 20-25.
95. P. J. Korytko, J. G. Scott; *Arch. Insect Biochem. Physiol.* **1998,** 37, 57-63.
96. J. F. Andersen; J. G. Utermohlen, R. Feyereisen; *Biochem.* **1994,** 33, 2171-2177.
97. J. F. Andersen; J. K. Walding; P. H. Evans; W. S. Bowers, F. Feyereisen; *Chem. Res. Toxicol.* **1997,** 10, 156-164.
98. J. G. Scott; *Insect Biochem. Molec. Biol.* **1996,** 26, 645-649.
99. J. G. Scott; M. Foroozesh; N. E. Hopkins; T. G. Alefantis, W. L. Alworth; *Pestic. Biochem. Physiol.* **2000,** (In press),
100. G. D. Wheelock; Y. Konno, J. G. Scott; *J. Biochem. Toxicol.* **1991,** 6, 239-246.
101. S. S. T. Lee, J. G. Scott; *Insect Biochem. Molec. Biol.* **1992,** 22, 699-711.
102. J. G. Scott, S. S. T. Lee; *Insect Biochem. Molec. Biol.* **1993,** 23, 729-738.
103. B. Lewin, *Genes VI;* Oxford University Press: Oxford, 1997; p. 1260.
104. F. J. Gonzalez, Y.-H. Lee; *FASEB Journal* **1996,** 10, 1112-1117.
105. J. G. Scott; N. Liu; Z. Wen; F. F. Smith; S. Kasai, C. E. Horak; *Gene* **1999,** 226, 347-353.
106. P. A. Zweidler-McKay; H. L. Grimes; M. M. Flubacher, P. N. Tsichlis; *Molec. Cell. Biol.* **1996,** 16, 4024-4034.
107. A. Kamat; J. L. Alcorn; C. Kunczt, C. R. Mendelson; *Molec. Endocrin.* **1998,** 12, 1764-1777.
108. H. Prapaipong; M. R. Berenbaum, M. A. Schuler; *Nucl. Acids Res.* **1994,** 22, 3210-3217.
109. C. Hung; R. Holzmacher; E. Connolly; M. Berenbaum, M. Schuler; *Proc. Natl. Acad. Sci. USA* **1996,** 93, 12200-12205.
110. P. A. Jones; *Trends Genet.* **1999,** 15, 34-37.
111. R. Holliday, G. W. Grigg; *Mutat. Res.* **1993,** 285, 61-67.
112. K. D. Robertson; S. D. Hayward; P. D. Ling; D. Samid, R. F. Ambinder; *Mol. Cell. Biol.* **1995,** 15, 6150-6159.
113. Y. Takahashi; C. Suzuki, T. Kamataki; *Biochem. Biophys. Res. Comm.* **1998,** 247, 383-386.
114. I. Vieira; M. Pasanen; H. Raunio, T. Cresteil; *Pharmacol. Toxic.* **1998,** 83, 183-187.
115. N. Yokomori; R. Kobayashi; R. Moore; T. Sueyoshi, M. Negishi; *Molec. Cell. Biol.* **1995,** 15, 5355-5362.
116. E. Matsunaga, F. J. Gonzalez; *DNA Cell. Biol.* **1990,** 9, 443-452.
117. G. Shaw, F. A. J.; *J. Biol. Chem.* **1992,** 267, 5515-5526.
118. W. Liang; J. He, A. Fulco; *J. Biol. Chem.* **1995,** 270, 4438-4450.

119. J. He, A. J. Fulco; *J. Biol. Chem.* **1991,** 266**,** 7864-7869.
120. G.-C. Shaw; C.-C. Sung; C.-H. Liu, C.-H. Lin; *J. Biol. Chem.* **1998,** 273**,** 7996-8002.
121. M. Cohen; J. Koener, R. Feyereisen; *Gene* **1994,** 146**,** 267-272.
122. T. Fournier; N. Mejdoubi; C. Lapoumeroulie; J. Hamelin; J. Elion; G. Durand, D. Porquet; *J. Biol. Chem.* **1994,** 269**,** 27175-27178.
123. P. Honkakoski, M. Negishi; *J. Biol. Chem.* **1997,** 272**,** 14943-14949.
124. E. Trottier; A. Belzil; C. Stoltz, A. Anderson; *Gene* **1995,** 158**,** 263-268.
125. Y. Park; H. Li, B. Kemper; *J. Biol. Chem.* **1996,** 271**,** 23725-23728.
126. S. S. T. Lee, J. G. Scott; *Pestic. Biochem. Physiol.* **1989,** 35**,** 1-10.
127. L. C. Terriere, In *Pest Resistance to Pesticides*; Georghiou, G. P. and Saito, T., Eds.; Plenum Press: New York, 1983; pp 265-298.
128. L. C. Terriere; *Ann. Rev. Entomol.* **1984,** 29**,** 71-88.
129. F. Carino; J. F. Koener; F. W. J. Plapp, R. Feyereisen, In *Molecular Mechanisms of Insecticide Resistance: Diversity Among Insects*; Mullin, C. A. and Scott, J. G., Eds.; Amer. Chem. Soc.: Washington D.C., 1992; Vol. 505 (ACS Symposium series), pp 31-40.

Chapter 3

Studies on Mechanisms of Insecticide Resistance in *Blattella germanica*

Fumio Matsumura[1,2], M. Miyazaki[1], K. Kaku[1], S. Inagaki[1], K. Ohyama[1], and D. Y. Dunlap[1]

[1]Departments of Environmental Toxicology and [2]Entomology, University of California, Davis, CA 95616

Blattella germanica has been a very good species for studying the mechanisms of resistance development. In this presentation we plan to describe our own efforts for elucidation of two cases of the resistance mechanisms in this species involving target insensitivities to two specific classes of insecticides: i.e. cyclodiene and *kdr* resistance. While the pattern of the shift of amino acids, resulting from mutation which specifically confer these resistance to *B. germanica* turned out to be identical to those simultaneously discovered in *Musca domestica* and *Drosophila melanogaster*, we found some differences in DNA base sequences and the nature of the sites specific mutations between *B. germanica* and these two dipteran species. One additional factor associated with the *kdr* phenomenon is the reduced mRNA expression of CaM-kinase. Since we could not find any mutation in the stretch of its DNA sequence which we could isolate, we conclude that this additional resistance factor is likely to be controlling the level of CaM-kinase expression.

© 2002 American Chemical Society

Introduction

Since there are several excellent review papers already published on the subject of mechanism of insect resistance due to "target-insensitivities", we do not plan to make this paper another review. Instead, we set our major objective to summarize the subject of resistance mechanisms in the German cockroach, *Blattella germanica*, an orthopteran species. The main reason is that most of works in this field have been carried out with *Musca domestica* and *Drosophila melanogaster*, two dipteran species, and we hope to offer the reader a comparative viewpoint. Another reason is that by re-analyzing and summarizing the vast amount of our own data, which we accumulated over the last 30 years in this laboratory, we hope to gain some perspectives on the meaning of what we have accomplished, what worked and what did not, and what kind of new questions we could raise for future research endeavors. At the same time, we hope to acknowledge key contributions made by other scientists as well.

With respect to new questions we raised, clearly the most important one is how we could explain the reason why a few mutations found in proteins constituting these ion-channels could confer such high levels of resistance to those insects. Since there is no detailed molecular description on these insect ion-channels available at this stage, all we could do in this paper was to offer our own speculation based on the recent models proposed by other scientists in other species. We constructed this model just to stimulate the interest of our colleagues. We had fun doing this and, hopefully, others will join in improving it or offering alternative models useful in resistance studies in the near future.

Background of discovery of resistance in *B. germanica*

Studies on the biochemical mechanism of insecticidal resistance of the German cockroach, *Blattella germanica* started around 40 years ago by Dr. A.W.A. Brown *(1)* of the University of Western Ontario, London, Canada. The original cyclodiene resistant strain, London strain was collected from an army camp, which had been repeatedly treated with probably chlordane. At the same time Dr. Brown obtained a strain *B. germanica* from Virginia Polytechnic Institute (VPI) which was found to be doubly resistant to DDT and cyclodiene. Along with several mosquito species which developed either cyclodiene-or DDT-resistance he and his associates during the period 1959 through 1962 characterized genetic modes of inheritance and established their resistance specificity to each class of chemicals, i.e. cyclodiene resistance extends its cross resistance to lindane but not to any other classes of insecticides, and DDT resistance is confined to DDT, pyrethroids and pyrethrins only. This was true even in the case of VPI strain, if one carefully selects individuals which showed no resistance to cyclodienes. The VPI substrain which showed specific resistance to DDT but not to cyclodienes was named as VPIDLS (DLS for dieldren susceptible). From VPIDLS strain, VT strain was procured through 8 generations of backcrossing to CSMA and continuous selection for DDT resistance. This concept of resistance specificity established at that time turned out to be very important, since, as we will eventually learn, that these two cases of resistance (i.e. one for cyclodiene and one for DDT) really represent two of the most clear cut examples of "target insensitivity-based resistance mechanisms".

These two standard resistance strains have been continuously maintained along with the standard susceptible strain, CSMA in the senior author's laboratory and over the years both resistant strains have been backcrossed for at least eight generations to create respective strains LPP and LT that maintained the respective original resistance gene and at the same time became congenic to the susceptible CSMA German cockroaches. Also from VT a substrain of VP was selected using permethrin as the selective agent. Before getting into the details of resistance mechanisms, we would like to describe briefly why we think *B. germanica* happens to become the species of choice for insecticide resistance studies for us. First, what we like most about *B. germanica* is the stability of the resistance phenotypes. While our experience is limited to the comparison of resistance patterns to houseflies and, to a lesser extent, pesticide resistant two-spotted spider mites (*Tetranychus urticae*) which were studied in most cases in parallel to *B. germanica* in our laboratory, the stability of resistance was always better in *B. germanica* as compared to *M. domestica* or *T. urticae*. In the case of houseflies the need for constant attention to maintaining the level of resistance through repeated selection and the avoidance of genetic drifts have been well known. There are other advantages of *B. germanica*. For instance, being hemimetabolous insects, cockroach nymphs show reasonably similar patterns of resistance to the adults unlike holometabolous dipteran insects such as houseflies and *Drosophila*, and, therefore, selection pressure could be applied at anytime even during the immature stage. At the same time, in the case of *B. germanica* the gender segregation is possible long before the adult emergence, making it possible to obtain resistant virgin females for controlled genetic experiments. Another feature we definitely took advantage of was their similarities to a well studied, and physically a larger species, *Periplaneta americana*, the American cockroaches. Thus, we could run some preliminary experiments including electrophysiological tests on the American cockroach first, and after miniaturizing the systems, we could run the same tests on *B. germanica*. Additional advantages worth mentioning briefly are: less likelihood of cross contamination (they are not flying insects), ease of rearing, less chance of chromosomal linking because of a large number of chromosomes they possess unlike the case of dipterans, and simply being an orthopteran species, offering the information on resistance mechanisms in non-dipteran species which have been the main targets of resistance mechanism studies by many insect toxicologists. Certainly, there are disadvantages of studying *B. germanica* as well, such as a lack of basic data and technologies (as compared to *Drosophila*), difficulties in relating the findings to others, limited genetic diversity, and probably low frequencies of mutations and therefore less likelihood of the presence of many resistance mechanisms to study a somewhat longer life cycle than houseflies and the absence of linkage markers. Nevertheless, in our case even some of these disadvantages gave us unique data and cross species perspectives which helped us to solve several key riddles in the field of insecticide toxicology as will be explained later.

Studies on the Pattern of Cross-Resistance and Preliminary Biochemical Studies

Since we already knew that these two cases of resistance in *B. germanica* involve "target insensitivity" mechanisms, our initial efforts were focused on "cross-resistance" studies using many toxic agents whose action mechanisms are well known. The rationale for this approach is that, if the target protein is specifically modified in the resistance insect so that the insecticide in question would not be able to exert its toxic action, that specific modification would confer cross resistance to agents which act directly on that target or act through that target protein to cause toxicity.

Following this basic game plan, we have tested many chemicals on cyclodiene resistant cockroaches first, but found that both London and Ft. Rucker strain (another cyclodiene resistant one from Texas) showed no cross resistance to many agents except to picrotoxin. The choice of this naturally occurring toxic chemical was based on the similarity of its chemical structure to some of cyclodienes which was noticed by the senior author during the seminar given by Dr. R. Beeman around 1978. We found that the level of cross resistance to picrotoxin was not only very significant, but also was "target-dependent", since we could demonstrate, through an electrophysiological method, that the abdominal nerve cord of the resistant *B. germanica* showed a significant level of cross-resistance to picrotoxinin, the toxic principle of the picrotoxin mixture (*2*) (Table 1). To test our hypothesis that the site of action of cyclodienes is identical to that of picrotoxin, as well as its principal active component picrotoxinin (i.e. $GABA_A$ receptor), we first studied the effect of cyclodienes on $^{36}Cl^-$ uptake *(3)*. This line of investigation was very fruitful in indicating the $GABA_A$ receptor to be the likely target for cyclodienes for the first time. The above finding, at the same time, indicated to us that this cross-resistance phenomenon is not merely coincidental, rather it is potentially related to the changes in their common target. This point was proven when Dr. Ghiassudin could show that there is a difference in ^{3}H-dihydropicrotoxinin binding affinities to nerve preparations from the susceptible and the resistant cockroaches (*4*). This binding assay approach was very laborious, but we wanted to use the same radiolabeled ligand as the original workers who proved the site of action of picrotoxinin to be the $GABA_A$ receptor (*5, 6*). Certainly the rest of story leading to the conclusion that the $GABA_A$ receptor particularly the picrotoxinin binding site is the main target of all cyclodienes and gamma-HCH (lindane) has been published and a number of review papers have already explained the details (*4-9*). Therefore, the only two points we wish to emphasize here are that during the initial period of investigation what kept our confidence high was the above observation of specific cross-resistance of these cockroaches to picrotoxin of which action on the $GABA_A$ receptor has been already well established. The second point is that the above example, along with other biochemical and physiological testes (*9, 10*), illustrates the usefulness of the German cockroach as the experimental animal, since most of techniques developed in the American cockroach could be applied to German cockroaches.

As for the case of DDT resistance, we ran a similar cross-resistance study *(11)* with different groups of agents with mostly known modes of action to affect the sodium channel (Table 2). In addition to DDT, early studies showed that only 4 compounds elicited definite cross-resistance in both DDT-strains of *B. germanica,* VT and VPIDLS. They were: grayanotoxin I, veratrin, EGTA and A23187. The first two agents are known to attack the sodium channel, so there was no big surprise there. Besides those two, aconitine, another sodium channel poison, also elicited a modest degree of cross-resistance in both strains. On the other

TABLE I. Minutes to onset of poisoning symptoms in the abdominal nerve cord of susceptible (CSMA) and resistant (LPP) German cockroaches

Experiment No.	Strains CSMA	LPP
Dieldrin (10^{-5}M)		
1	41	65
2	57	87
3	71	77
4	57	>140[a]
5	43	> 90[a]
($\bar{X} \pm$ SE)	53.8 ± 5.5	> 91.8 ± 12.5
Picrotoxinin (10^{-5}M)		
1	41	>125[a]
2	46	>135[a]
3	91	>135[a]
4	69	> 86[a]
($\bar{X} \pm$ S.E.M.)	61.8 ± 11.5	>120.3 ± 11.7

[a]Maximum time period observed at which time the experiment was terminated.

[b]Significant difference at $P \leq 0.05$ as judged by sign test (5).

TABLE II. Susceptibility levels (24hr LD_{50}) of resistant and susceptible strains of German cockroach against various neuroactive agents* (14).

	Cockroach strains				
	CSMA	VT	Ratio† (resistance)	VPIDLS	Ratio† (resistance)
Insecticides					
DDT	60	121	(2.0)	320	(5.3)
Diazinion	0.25	0.30	(1.2)	-	-
Nicotine	37	36	(0.97)	-	-
Carbaryl	18‡	14‡	(0.78)	21‡	(1.2)
Agents affecting Na^+, K^+ Channel					
Grayanotoxin I	18	48	(2.7)	80	(4.4)
Valinomycin	0.44	0.50	(1.1)	1.8	(4.0)
Aconitine	6.6	10	(1.5)	16	(2.4)
Veratrin	3.4	7.1	(2.1)	11	(3.2)
Calcium§ modulator					
Calmidazolium	44	39	(0.88)	108	(2.5)
TPZ‖	64	96	(1.5)	179	(2.8)
Chlorpromazine	24	17	(0.71)	92	(3.9)
A23187•	1.3	6.4	(4.8)	13	(9.4)
Gramicidin D	3.1	5.4	(1.8)	8.6	(2.8)
EGTA	91	206	(2.3)	291	(3.2)
Lanthanum	65	66	(1.1)	73	(1.1)

*Data are expressed in terms of dose needed to kill 50% of population. All cockroaches were treated with 30μg/cockroach of piperonyl butoxide (topical) prior to the test.

†Resistance ratio between LD_{50} of resistant (VT or VPIDLS) divided by LD_{50} of susceptible (CSMA) strain. A value less than 1.0 indicates that the suceptible strain is more resistant.

‡LT_{50} in hr with piperonyl butoxide.

§Other compounds to which VT did not show any significant cross-resistance were: D600, nitrindipine, verapamil and PCMPS.

‖TPZ (=trifluoroperazine) EGTA (ethylene glycol-bis (*p*-aminoethyl) ether N,N,N,N-tetraacetic acid).

•Calcium ionophore.

hand, significant cross-resistance to the last two agents elicited was quite a surprise. A23187 is a well established Ca^{2+}-selective ionophore and EGTA is known to specifically chelate Ca^{2+} over any other divalent cations. Since their main action mechanisms are very well acknowledged, and since the tendency was consistent even in VT strain which has been made to be congenic to CSMA, the inevitable conclusion we had to reach was that in addition to the sodium channel a calcium regulation process (or processes) was involved somewhere in the development of DDT resistance in *B. germanica.* Another type of cross-resistance studies we conducted was those for pyrethroids. Two studies by Scott and Matsumura (*12, 13*) showed that these DDT-resistant German cockroaches are clearly cross-resistant to pyrethroids, particularly to type I pyrethroids as judged by both *in vivo* mortality tests and *in vitro* electrophysiological tests. The levels of cross-resistance to pyrethroids were found much higher when knockdown was used as the criterion rather than those obtained by the standard mortality testing. Based on those observations we concluded that it would be appropriate to call the case of DDT-resistance in strains of VPI lineage to be *kdr* resistance in analogy with the case in houseflies. Although at that time some objections were raised, as will be shown later from the molecular biological point of view, this turned out to be a correct decision to make.

We have also taken full advantage of the similarity between German cockroaches to American cockroaches and run a number of electrophysiological (*12, 13*) and biochemical experiments using synaptosomal preparations from the former. One example is the tetrodotoxin-sensitive Na^{22}-uptake experiment (*14*) where we could show that the stimulating effect of DDT on this process is much more pronounced in synaptosomal preparation from the susceptible CSMA cockroaches than that from the DDT-resistant cockroaches (Table 3), confirming the site of alteration for resistance to be the sodium channel.

Molecular Investigations on the GABA Receptor of *B. germanica*

In 1993 this laboratory published *(15)* that a GABA receptor subunit, at that time we called β subunit, isolated from cyclodiene resistant German cockroaches (LPP strain) showed a consistent shift in one amino acid alanine (A) to serine (S) (Fig. 1). While we found some additional mutations in other locations in at least two of β subunits, resulting in a shift in amino acid, we considered the above A to S shift to be the one covering the expression of cyclodiene resistance by the

TABLE III. Effect of DDT and TTX on $^{22}Na^+$ uptake activity of synaptosome preparation from DDT-susceptible (CSMA) and – resistant strains (VPIDLS and VT) of the German cockroach (14).

	$^{22}Na^+$ uptake nmole/mg 30 sect		
Treatments*	CSMA	VPIDLS	VT
Control	0.242+0.038	0.230+0.033	0.230+0.014
+DDT	0.451+0.095	0.264+0.117	0.237+0.021
+TTX	0.049+0.062	0.115+0.092	0.131+0.060
+DDT +TTX	0.175+0.035	0.196+0.035	0.119+0.013

*Concentrations were: DDT 10^{-5}M and TTX 10^{-6}M.

†Data expressed as mean ± SE 3-6 determinations.

Species																											
German cockroach (susceptible strain)	V	S	F	W	L	N	R	N	A	T	P	A	R	V	A	L	G	V	T	T	V	L	T	M	T	T
German cockroach (resistant strain)	V	S	F	W	L	N	R	N	A	T	P	A	R	V	**S**	L	G	V	T	T	V	L	T	M	T	T
Drosophila	V	S	F	W	L	N	R	N	A	T	P	A	R	V	A	L	G	V	T	T	V	L	T	M	T	T
Lymncea β	V	S	F	W	I	N	H	E	A	T	S	A	R	V	A	L	G	I	T	T	V	L	T	M	T	T
Mammal (β subunit)*	V	S	F	W	I	N	Y	D	A	S	A	A	R	V	A	L	G	I	T	T	V	L	T	M	T	T
Mammal (α subunit)†	V	S	F	W	L	N	R	E	S	V	P	A	R	T	V	F	G	V	T	T	V	L	T	M	T	T
Mammal (γ2 subunit)‡	V	S	F	W	I	N	K	D	A	V	P	A	R	T	S	L	G	I	T	T	V	L	T	M	T	T
Rat brain (δ subunit)	V	S	F	W	I	S	Q	A	A	V	P	A	R	V	S	L	G	I	T	T	V	L	T	M	T	T
Human (ρ1 , rho1 subunit)	V	S	F	W	I	D	R	R	A	V	P	A	R	V	P	L	G	I	T	T	V	L	T	M	T	T
Human (rho2 subunit)	V	S	F	W	I	D	R	R	A	V	P	A	R	V	S	L	G	I	T	T	V	L	T	M	T	T

Figure 1. Comparison of the amino acid sequences of the area of GABA receptor containing the mutation resulting in the cyclodiene resistance in *B. germanica*. The location of the mutation in the resistant strain is indicated by a bold letter. From (15) with the permission of the Pesticide Science Society of Japan.

following reasons: first, this shift occurred in one of the most conserved region; second, the same amino acid shift (A to S) was observed in the cyclodiene resistant *Tribolium castenium* although the nature of DNA sequence and the base involved in this mutation were very different (Fig. 2), third, it is located in the center of the inner wall of the proposed chloride channel, and fourth, this mutation was found in all of the 4 variants of β subunits found in the cyclodine-resistant *B. germanica*.

In the same year ffrench-Constant et al. (*16*) and Thompson et al. (17) reported the equivalent mutation of A to S in the corresponding area of the GABA receptor of *Drosophila* and *Aedes aegypti*, respectively. We were relieved to know that our finding coincided with theirs, providing an instant confirmation that our judgement was correct. At the same time, we found it very interesting that this particular mutation and the M2 region surrounding this site was almost identical between *B. germanica* (*18*) and *Drosophila* (*16*) subunits, despite the fact that there are many differences in the sequence between these two. Perhaps the most significant difference was in the length of the amino acid sequence of M3-M4 region. The one from *Drosophila* had 230 amino acids whereas that of *B. germanica* had only 75 amino acids (*16*). Not only that, there were no similarities in amino acid sequence in this region between these two. In addition, we wondered why in *B. germanica* there are 4 variant subunits (we now call them Rdl, 1, 2, 3, 4, since the sequence of the susceptible (wild) type of the corresponding Drosophila subunit was sequenced earlier by ffrench-Constant et al. (*19*)). While we cannot answer this question until we know much more about the organization of GABA receptors in insects, judging by the fact that all of these had the same A to S shift, we consider that all of 4 variants are likely to be participating in the expression of resistance and hence in the maintenance of their normal functions.

For the purpose of this chapter, what we would like to stress next is the importance of accumulation of the background biological and toxicological data in the same species in interpreting such a molecular biological finding back to the meaning of the differential expression of toxic effects. The main questions we must address are (a) why such a small, single amino acid change confers the specific resistance to cyclodienes, and (b) whether or not this is the site of attack by all these cyclodiene insecticides. Our current opinion, which still needs experimental confirmation, is that this alanine site itself is indeed the site of binding of all of cyclodiene type insecticides and, therefore, the reason why this shift gives the resistant cockroach the survival advantage is that the switch of alanine to a more polar amino acid serine (addition of –OH)

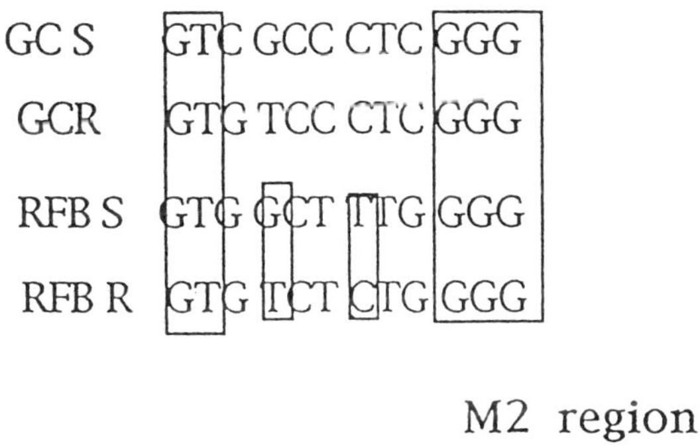

Figure 2. Comparison of DNA sequences of a lower M2 region of the GABA receptors of two of cyclodiene-resistant species, B. germanica and Tribolium castenium (red flour beetle). Note that in both cases mutations resulted in an alanine (A) to serine (S) shift. From (6) with the permission of the Pesticide Science Society of Japan.

at this hydrophobic site makes it more difficult for these insecticides to bind to the inner wall of the channel. There are reasons why we consider the above possibility being likely. As early as 1966 Dr. Hayashi in our laboratory (*20*) showed that the nerve membranes from the cyclodiene resistant LPP bind less with ^{14}C-dieldrin than did the same preparations from the susceptible CSMA cockroaches. Furthermore, Telford and Matsumura (*21*) ran autoradiogram of ^{14}C-dieldrin bound intact cockroach abdominal cords at an electroumicroscopic level and found that the number of silver grains, representing ^{14}C-dieldrin, attached to the membranes of the axonic and the synaptic region was much less in number in the preparations from the cyclodiene resistant German cockroaches (London strain) than those from the susceptible individuals. Later, Tanaka and Matsumura (*22*) and Deng et al. (*23*) have also shown the same tendency in houseflies. Thus, the preponderance of evidence points to the likely scenario that this site of mutation itself is the actual site of attack by cyclodienes type insecticides which results in the interference of the chloride channel opening mechanism. As will be explained later, the fact that such a mutation-induced shift in the amino acid composition occurred right on the surface of the inner wall of the chloride channel proper makes it easier to accept this most simplistic explanation of the event.

Molecular Basis of *kdr* Resistance and Its Implications for the Sodium Channel Operation

As we briefly mentioned already the *kdr* resistant German cockroaches show cross-resistance to pyrethroids, particularly to type I pyrethroids, and somewhat less cross-resistance, to type II pyrethroids. This tendency was more pronounced when we adopted the toxic endpoint of knockdown, rather than mortality, as a criterion, or the surface contract method as opposed to topical application method, as the test approach (*24*). Since electrophysiological evidence in this species (*12*) as well as toxicological observations (*14*) have already indicated that the nervous system of *B. germanica* responds quite different way to type I from the way they react to type II pyrethroids, we speculated that the above cross-resistance pattern, favoring type I pyrethroids over type II, is likely representing the basic difference between these two groups of pyrethroids in influencing the sodium channel and the effectiveness of the *kdr* modification on the sodium channel in counterbalancing the action of these pyrethroids rather than a simple reduction in the number of the sodium channel, or the existence of several types of sodium channels. We further speculated that the *kdr* mechanism in *B. germanica* is very similar to that in *M. domestica* based on the similarities in pyrethroid resistance patterns.

This prediction turned out to be true. By 1995 we could finally obtain DNA strands encoding the entire membrane spanning region of para sodium channel gene from both susceptible and resistant *B. germanica* and sequenced (*26*). The complete data on the entire sequence of IS1 through IVS6 has been deposited in GenBank. When we examined both *B. germanica* strains side by side, it became apparent that there is a mutation at the lower center of the IIS6 membrane spanning segment which induced a shift in amino acid composition at this site to phenylalanine from the original amino acid leucine. To make sure that this mutation is the one causing the *kdr* type resistance, we worked on the housefly *kdr* strain along with the susceptible SBO strain using a similar approach but with different PCR primers, and found that the *kdr* strain has the same L to F shift at the equivalent site in IIS6 segment as well (Fig. 3). It is clear that this particular lower region of the IIS6 segment is very much conserved, and the only sodium channel sequence other that those two *kdr* strains showed phenylalanine instead of leucine in this position was the *Drosophila* DSC sequence. While the function of DSC type sodium channel in insects has not really been elucidated, as judged by the frequent detection of this gene product by the use of RT-PCR, this gene product must be also abundant in *B. germanica*. Therefore, it is likely that DSC is also a functional sodium channel in some capacity. The conclusion one can reach from such a consideration is that this L to F shift in para sodium channel is not likely to totally impede the functionality of the sodium channel.

Our reporting on the elucidation of the *kdr* mutation site coincided with that of Dr. Devonshire's group on *kdr* and *Skdr* of houseflies (*27*) in the same journal. Again, it is really pleasing to know that the nature of this site specific "target insensitivity" based DDT-resistance mechanism is identical in more than one species, offering an instant confirmation. These findings were soon followed by other groups' reports (*28, 29*) on the same mutation at the same site of the insect *para* sodium channel as in these two species of insects.

So what is the mechanistic meaning of this specific shift of amino acid leucine (L) to phenylalanine (F) at this site? Before getting into this discussion, there is one more important piece of evidence to consider. That is, there are now several cases of "target modification" found in DDT/ pyrethroid resistant insects as will be shown later. For instance, Pittendrigh et al. (*30*) have characterized mutational changes in the para-type sodium channel in six laboratory selected DDT/pyrethroid

Species	Channel	Sequence
B. germanica	(para CSMA strain)	W S C I P F F L A T V V I G N L V V L N L F L A L L L S N
B. germanica	(para VT strain)	W S C I P F F L A T V V I G N F V V L N L F L A L L L S N
M. domestica	(para SBO strain)	S S C I P L L L A T V V I G N L V V L N L F L A L L L S N
M. domestica	(para KDR strain)	S S C I P L L L A T V V I G N F V V L N L F L A L L L S N
Drosophila	para	V S C I P F F L A T V V I G N L V V L N L F L A L L L S N
Drosophila	DSC-1	L T C F A I F L P T L V M G N F M V L N L F L A L L L N S
Rat Brain I	RSC-1	A M C L T V F M M V M V I R N L V V L N L F L A L L L S S
Rat Brain II	RSC-2	T M C L T V F M M V M V I G N L V V L N L F L A L L L S S
Rat Brain III	RSC-3	T M C L I V F M L V M V I G N L V V L N L F L A L L L S S
Rat Muscle	RSKM-1	A M C L T V F L M V M V I G N L V V L N L F L A L L L S S
Rat Muscle	RSKM-2	S L C L L V F L L V M V I G N L V V L N L F L A L L L S S
Rat Heart	RH-1	S L C L L V F L L V M V I G N L V V L N L F L A L L L S S
Human Brain	HBSC-1	T M C L T V F M M V M V I G N L V V L N L F L A L L L S S
Human Heart	HH-1	S L C L L V F L L V M V I G N L V V L N L F L A L L L S S
Human Muscle	HSKM-1	A M C L T V F L M V M V I G N L V V L N L F L A L L L S S
Human Heart and Uterus	HNaVZ-1	S W C I P F Y L M V I L I G N L L V L Y L F L A L - V S S
Eel		P M C L A V Y M M V I I I G N L V M L N L F L A L L L S S

Figure 3. Comparison of amino acid sequences of IIS6 regions of sodium channels from various species and tissues. The kdr resistant strain is shown on the second row in comparison to the susceptible strain on the top row. The L to F shift is the one conferring *kdr* resistance which was also found in the kdr resistant *M. domestica* as shown in row 3 and 4. Note that this area is well conserved except the case of DSC-1 of *Drosophila*. From (26) with the permission of Springer-Verlag, Inc.

resistant strain of *Drosophila* and found that there are four sites of mutations that could be related to 10- to 30-fold increase in DDT–resistance. They are domain I between S4 and S5, domain III between S4 and S5, domain III within the pore region III (on III 5-6 linker) and domain III within S6, in addition to the same *kdr* and *skdr* sites of mutations (also see later discussions). Their study points out the existence of several "target insensitivity" sites at least in the case of laboratory selected *Drosophila melanogaster* populations. Their finding is important from the viewpoint of providing the clue to the molecular basis of the action of DDT and pyrethroids on the sodium channel. It must be noted also that two of the mutations involve the *kdr* site on IIS6 and a site on IIIS6 (their designation para 74) three mutations occurred on the super *kdr* site and equivalent sites on the IS4-IS5 and IIIS4-S5 linkers. One additional mutation on the IIIS5-6 linker itself did not match with any of the previously reported *kdr* or *skdr* mutation sites. The location on III5-6 linker coincides with the segment considered to be involved in the inactivation process of the sodium channel. It may be speculated that those equivalent site on IS5-6 as well as the *Skdr* sites may also participate in a similar function. The functional meaning of the *kdr* site and the similar para 74 site (a shift of methionine to isoleucine) is not known.

To understand the functional meaning of the kdr mutation on the sodium channel, it is necessary that we critically study the known action models which have been proposed for sodium channels as well as for other types of ion-channels. The model of sodium channel frequently sited is that of Sato and Matsumoto (*31*). In most models prior to that model both IIS6 and IIIS-6 had been considered to be serving merely as the rigid outer wall supporting the scaffold for the inner channel composing segments such as S2's and S4's. However, in the model proposed by Sato and Matsumoto (*31*) S6 segments have been proposed to serve as the base for the functionally important S5-6 linkers which are likely to be the segments envisioned to slide in and out of the channel itself. They also are known to serve as an anchor base for the important II-III linker which participates in the opening and closing of the cytosolic side of the gating, coordinating the movement of the bulky, negatively charged C-terminus protection mass as well as the positively charged III-IV linker (Fig. 4).

However, more information is needed to understand the mechanism by which the L to F amino acid shift in the IIS6 outer segment induces *kdr* resistance. Here, we will attempt to explain a possible mechanism through which such a mutated form of the sodium channel could resist the action of DDT and pyrethroids by assuming that

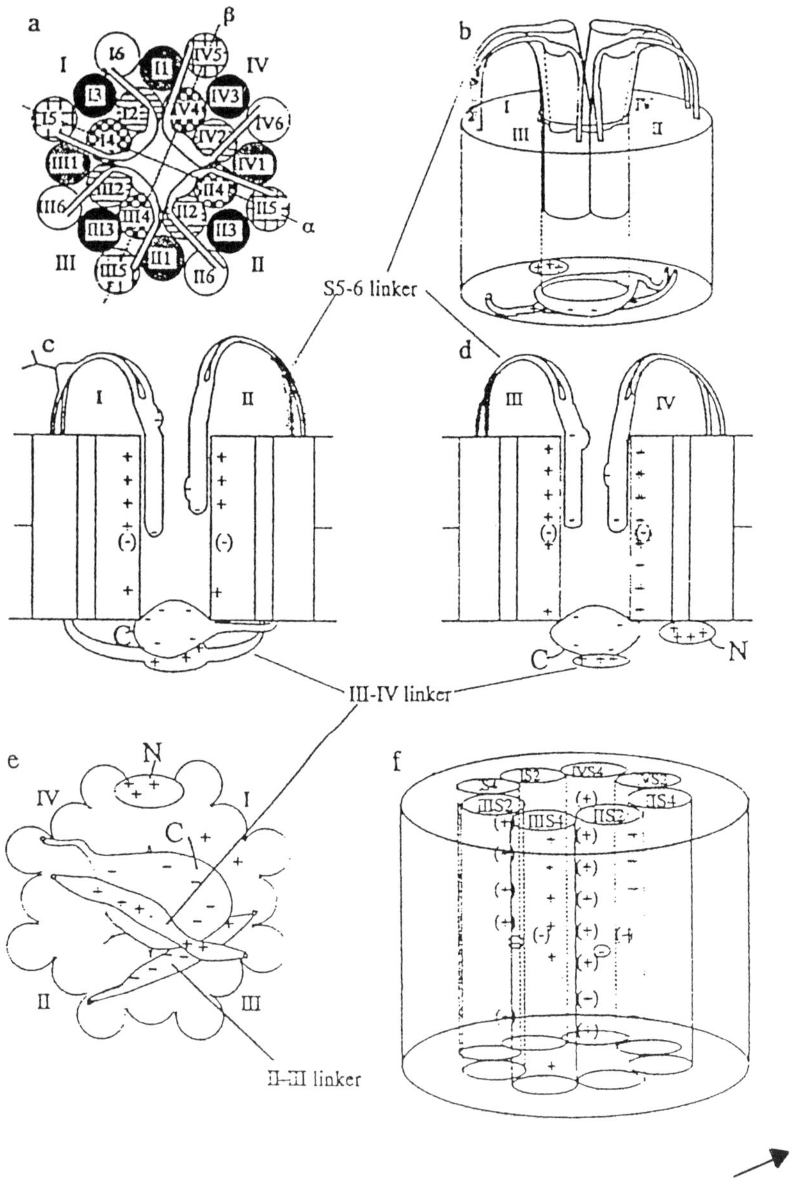
a
b
c
d
e
f
S5-6 linker
III-IV linker
II-III linker
β
α
N
C

Figure 4. Schematic illustration of the sodium channel operation model proposed by Sato and Matsumoto (31), indicating its tertiary structure state of (a) resting, (b) initially activated, (c) fully activated, (d) inactivated, (e) restoring resting status. Note that in this model the most prominent features are (c) the negatively charged S5-6 linkers (the arching structure with flaps) sliding into the channel to shield the series of the positively charges if the S4 segments comprising the inner surface of the channels (ii) the important role of the negatively charged C-terminal moving away as the channel opens and its temporary attachment to the positively charge N-terminal, (iii) the roles of the positively charged III-IV linker guiding the movement of the C-terminal and affecting the shape of the cytoplasmic opening of the channel, (iv) the action of the II-III linker to start the process of inactivation by pulling back the III-IV linker to the original position, and (v) the final movement of the C-terminal to plug the cytoplasmic and of the channel. Directly taken from Sato and Matsumoto (31) with the permission of Springer-Verlag, Inc., New York.

basic similarities must exist among operations of ion-channels. The most recent interpretation the ion gating mechanism based on the acetylcholine receptor (*32*) and potassium channel (*33*) is that the opening of the channel involved alternative association of helix lining the inner wall of the channel involving movements of whole subunits rather than of single domains or segments of the channel (*33, 34*). If we can assume that the operation of the sodium channel is similar to those (i.e. K+ channel operation) the helix of inner wall segment proteins (S2 and 4's) are likely to interact with the segments constituting the outer wall (S5 and 6's) through rearrangements and rotation of the individual helix conformations. The wild form of the para proteins allow DDT to interfere with to induce these conformational changes, but the *kdr* form would not let these insecticides to interfere or it would compensate the channel to counteract the end result of DDT's action. Whether the site of *kdr* mutations/ amino acid shift is also the site of DDT binding is the question we must address. There are some indications that less amounts of DDT bind to the nervous system of the *kdr* insects (*35*), but those classic study results are not so definitive. Besides, there are cases of DDT resistant insect nerves possessing quantitatively less numbers of sodium channels. If this amino acid shift is also the site of DDT and pyrethroid binding, one must come up with a logical explanation why such a shift from leucine to a phenylalanin (or methionine to isoleucine or any other shifts) could reduce the binding affinity to these insecticides.

When one considers the fact that there are now several sites of mutations/amino acid shifts in the molecule of the sodium channel protein, it is more logical to consider that these changes offer remedial responses or counter acting strategies to compensate for the damages caused by DDT and pyrethroids by the resistant insects so that the sodium channel would continue to operate even though these insecticides continue to modify the channel. The main reason for the support for this "compensatory change" theory is the fact of multiple sites of mutations. Indeed, it is difficult to conceive that these chemicals bind to so many different sites and still cause qualitatively identical symptoms. The second reason why we think "compensatory change" theory is a more plausible explanation is that *kdr* resistance in *B. germanica* confers cross-resistance to grayanotoxin I, veratrin as well as to aconitine. Of these, grayanotoxin I and veratridine/batrachotoxin are known to bind to the positive change of III and IVS4 segment on the inner channel wall very close to the cytoplasmic end (i.e. 24th residue) (*31*), a totally different site from the *kdr* mutation site. Yet, it is clear that the *kdr* mutation confers cross-resistances to these sodium channel opening toxins binding to a different site. Thus, the most logical way to explain this cross-resistance phenomenon is that this *kdr*-causing shift in amino acid must somehow bring the conformational change to make the sodium channel operates reasonably well despite the influences of these channel poisons. The third reason why we think this is the better way to explain the observed phenomenon is the advancement in the understanding of the molecular basis of ion channel operations.

Until quite recently, most scientists envisioned the operation of ion-channels to be carried out by simple gating mechanisms. In the case of the sodium channel, the existence of both the external gate and the internal gate have been considered as both theoretical and physical entities. Indeed the negatively changed carboxy terminal mass is still considered to plug the cytoplasmic opening of the channel at the return of the channel to the resting state after the membrane repolarization. Additionally, the physical interactions between the highly positively changed S4 subunits, lining the inner wall of the channel, and the negatively charged S5-6 linkers (to sequester or cover up the positive changes of S4) are the likely physical events triggering the entry of sodium ions into the channel, representing the hypothetical opening of the outer gate (*31*). Nevertheless, the most recent study results indicate that the ion channel operations require much more complex conformational changes involving the entire membrane spanning segments/helices as well as coordinated movements of the entire subunit including helices and connecting linkers, rather than movements of single domains, C-terminal protein mass alone or individual linkers (*34*). Furthermore, the whole motion of opening and closing of ion channels apparently requires twisting of the bundles of several membrane spanning helices (segments) lining the channel proper so that some of those would depart from the upright position perpendicular to the lipid bilayer to assume bent and/or tilted position. A good example in the case of the ion channel of the nicotinic acetylcholine receptor (*32*). This gated ion channel has five subunits that constitute the lining of the channel. Upon the arrival of ACh the channel opens which involves substantial rearrangements so that those channel lining helices to assume an alternative association (see Fig. 5). Note that the position of those helices substantially deviating from the perpendicular orientation as well as the bend (i.e. the "kink") of the helices that are pointing to the center of the pore at the "closed" state. Upon the arrival of acetylcholine these helices start rotating 90° counterclockwise in unison to create a larger circumference of the narrowest passage to facilitate the greater twisting of the whole bundle of individual segments.

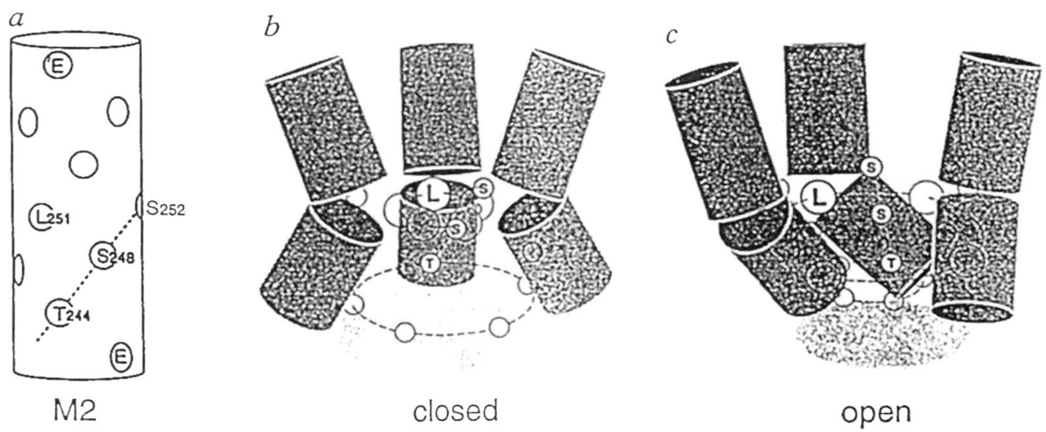

Figure 5. A simplified model illustrating general movement of pore-lining segment M2 of the acetylcholine receptor to create the open state (on the right) from the closed state upon the arrival of acetylcholine, viewed from the inside (one M2 segment directly on the front has been removed to clearly show the changes in other three segments). In this scheme these inner segments twist counterclock wise to open. The kink (bends) is created by bulk leucine (L) and three small and polar amino acids (serine and threonine) (see (a) left). This bent helix shape is very important, since the twist of the entire channel, coupled with the rotation of individual segments create the wider central pore for opening by moving the protruded bends of leucine away from the position of pointing at the center of the pore to the parallel orientation along with the circular shape of the outer wall. From Urwin (32) with the permission of Nature, New York office.

While nobody knows the precise molecular movements of complex ion-channels such as the para-type sodium channel which is operated by 24 complete individual segment helices plus a number of interconnecting linkers among segments and domains in a coordinated manner in the channel opening and closing motions, we must make the necessary effort to incorporate the best available data in our analysis of the meaning of *kdr*-type mutations on the sodium channel operations. The best available information to date is on the operational mechanism on the KcSA potassium channel of streptonyces lividans for which chrystal structure analysis data are available (*36*). The most recent study by the same research team (*33*) of the channel gating movements based on the site-directed spin-labeling and electron paramagnetic resonance spectroscopic analyses on this channel revealed that the twisting of the whole bundle of subunits is also coupled to the rotational movement of some key subunits (see Fig. 6) as well.

This KcSA potassium channel proper consists of 8 helices, 4 TM1 (residues 26 to 50) and 4 TM2 (residues 90 to 120). The inner wall of the channel is lined with 4 TM2 helices and the outside wall is comprised with 4 TM1 helices. The longer TM2 helices are tilted so that their cytoplasmic ends are much closer to each other than the extracellular ends, forming an upright funnel shape. At the time of the channel opening the whole bundle as well as individual TM2's rotate counter clock wise to make the diameter of ion permeation pathway longer (Fig. 6). The extent of the movement is larger at the cytoplasmic end. The most interesting part of their movements for us is that the counter clock rotation of the individual subunits is also observed among TM1 helices that comprise the outer wall. These changes are coupled with a small degree of tilt of TM1's toward the center, perhaps to fill the space vacated by the outwardly expanding TM2 helices. The reason why we are interested in this particular phenomenon is that the *kdr* mutation in *B. germanica* and *M. domestica* occurred on the lower center of IIS6 which makes up the outer wall as one of surrounding helices. That means the helices comprising the outer wall of this type of ion channels may not be mere immobile structure or physical scaffolds. In the case of the KcSA potassium channel, at least, they are rather active participants of gating operations. There is already a good example of the importance of an outer wall segment component in determining the sodium channel inactivation mechanism. That is, a substitution of one amino acid in the IIS5 segment is known to cause hyperkalemic periodic paralysis (HYPP), human disease marked by phenotypic deficiency in sodium channel inactivation operation (*37*). The most intriguing aspect of this T to M mutation took place on the equivalent threonine site of mutation on IIS5

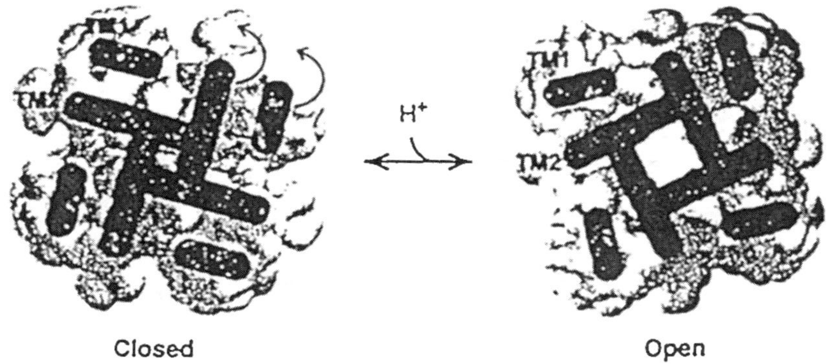

Figure 6. The molecular mechanism of activation gating of KcSA, K^+-channel. The model is viewed from the extracellular side. Four TM_1 segments comprise the outerwall and four TM_2 the inner pore. The opening of this K^+-channel is triggered by a proton-dependent mechanism. Note that both individual TM_1 and TM_2 must rotate counterclockwise to create a larger pore opening. Although the movement of TM_1 mirrors the rotation of TM_2 but the former helices tilts more inward toward the permeation pathway, probably to help the movement of TM_2. Taken from (33) with the permission of the American Association of the Advancement of Science.

which involved a T to I shift and gave DDT/pyrethroids resistance to head lice (*38*) and diamondback moth (*39*). The above HYPP mutation caused a shift in activation toward hyperpolarization and dramatically impaired slow activation of the muscle sodium channel linker. Another HYPP mutation has been reported to occur on the lower middle point of IVS6, another outer wall segment. This I to F shift occurs at an equivalent location of the kdr site on IIS6. This mutation also shifted both steady-state activation and inactivation to hyperpolarizing direction (*37*).

Regarding the *kdr* mutation on IIS6, the site of this L to F shift occurred at the 8th amino acid counting from the cytoplasmic end approximately 1/3 at the lower makes of the total (23 amino acids) IIS6 segment. According to the model of Sato and Matsumoto (*31*), the most closely associated neighboring segment facing the channel inner wall is IIS2. It must be noted that this mutation site coincides with the position of the central negative change (glutamic acid) of the IIS2 segment which likely effects the voltage sensitivity (*31*) as well as the channel closing to restore the resting state at repolarization. This idea is interesting, since the rotational changes of segments such as the one occurring in the conformational changes of K+ channel are at least conceptually similar to this proposed sequestration of the negative charge of IIS2 from the open face of the inner wall of the Na-channel. Considered from this viewpoint the L to F shift of amino acid observed in permethrin-resistant head lice (*38*) on IIS5 (at position #932) occurred at an equivalent position (the 7th amino acid from the cytoplasmic end). On the other hand, the position of the M to I shift occurred in the para74 mutant of DDT and pyrethroid resistant *Drosophila* is at the 3rd amino acid from the extracellular end of the IIIS6 segment (Fig. 7). Another mutation causing kdr-like resistance was found to occur on IS6 segment (V to M at 421) at about an equivalent site of *kdr* mutation (*40*). IS6 is another segment constituting the outer wall. Thus, the amino acid switches involving the outer wall segments to make the sodium channel more refractory to these insecticides occur at several locations of amino acid and heights, but the important consideration must be that all those changes must somehow countcract against the action of these insecticides to make the process of inactivation and subsequent restoration of the resting state of the channel difficult. One interesting observation is that, in the case of the acetylcholine receptor studied by Unwin (*32*), the centrally located leucine has been determined to provide the "kink" to the helix. These "kinks" offer the means to increase the power of the rotational effect to close and open the channel. The helices of the sodium channels have centrally located leucines as well as

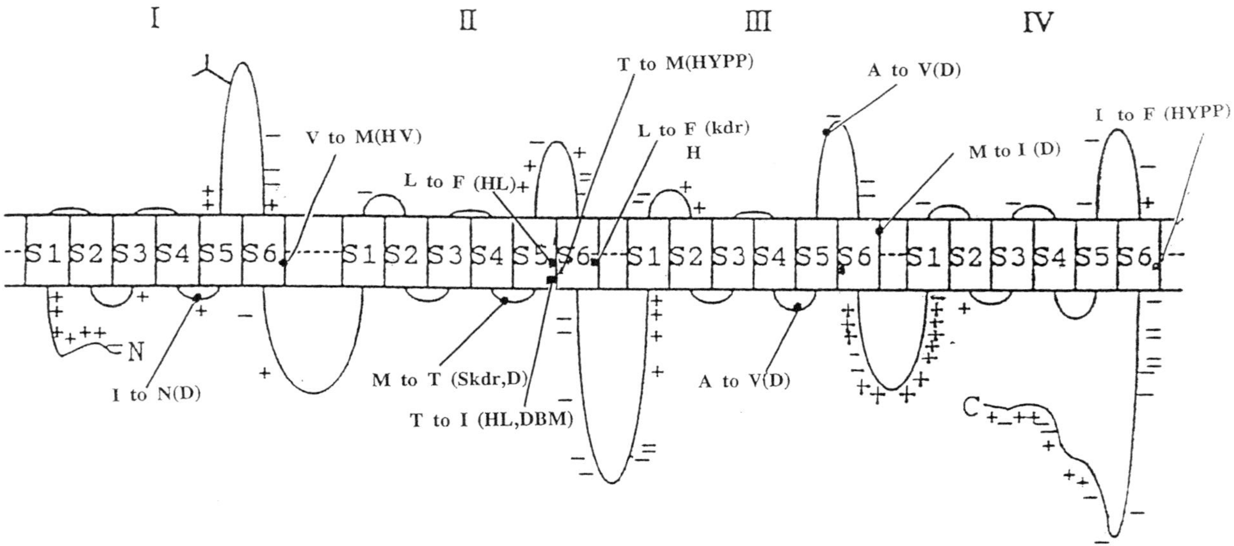

Figure 7. A schematic diagram of the sodium channel organization indicating the approximate location of mutations resulting in the shift of amino acids and expression of DDT/pyrethroid resistance Domain numbers are shown as I, II, III and IV and segment designations are given by S1,2,3,4,5 and 6. The abbreviations given are D for *Drosophila* (30), HL for headlice (38) DBM for diamondback moth (39). *Skdr* for houseflies (27), *kdr* for *B. germanica* (26,29) houseflies (26,27,46) and *Anopheles gambiae* (44), *Haematobia irritans* (43) and *Plutella xylostella* (diamondback moth) (39) etc. HYPP represents an abbreviation of human hyperkelamic periodic paralysis, a disease caused by inactivation deficiency of the sodium channel (37 also see text). Modified from the orginal figure of Sato and Matsumoto (31) with the permission of Springer-Verlag, Inc., New York.

adjacent small amino acids to provide kinks. These observations favor the idea that the *kdr* shift is providing a kinetic shift toward favoring the easy closure of the channel. Another phenomenon we noticed among all these mutations involving the outer wall segments is that they cause shifts of amino acids to become bulkier in size and in some cases involving IIS5 and IIS6 less polar. These observations support the view that the cytoplasmic half of the IIS5 and IIS6 and probably IS6 segments are very important in achieving proper conformational changes involved in inactivating and restoring the resting state.

Is there any other model than the above one proposed by Sato and Matsumoto (*31*) which could shed new insight on the role of IIS6 segment in the operation of the sodium channel? In this regard, the most recent 3-deminsional topographic characterization of the sodium channel from the elective eel, through cryo-electron microscopic visualization is most interesting (*41*). In this article, Sato et al. (*41*) report remarkable features of the sodium channel which have not been considered before. The first surprising feature is, that in addition to the central chamber (i.e., the main ion-channel), they propose that there are 4 side channels which are joining the central channel's opening near but below the outer surface so that there are 4 joint openings at the outer surface. Those same 4 peripherally located side channels, however, directly open to the inner surface (cytoplasmic side), though each opening will be very close to the main channel opening, so that altogether 8 openings are found on the cytoplasmic side. The benefit of this most unexpected feature of this new model, based on 3-dimensional visualization, is that it allows scientists to postulate a totally new way of interpretation on the operational modes of the channel (*42*). The most likely scenario is that these side channels are mainly operated by the higly charged S4 segments which are known to move up and down during the gating operation. This movement cannot be just linear (a straight up and down motion), since the topographic picture shows a narrow constriction at the center of the main channel with a top cover consisting of four 5-6 linkers. Thus, the likely scenario is that the movement of S4 segments requires additional twisting and turning movements as in the case of ACh receptor to achieve their up and down translocation. Another interesting scheme postulated by Catterall (*41*), who examined this new model in detail, is that the inner surface of the main chamber of the sodium channel is likely to consist of S6 and S5. If such is the case, the site of *kdr* mutations in the German cockroach sodium channel corresponds to be just below the narrow "constricted" passage point within the main chamber (visualize an hourglass shape with the cental pore where the speed of the passage of sand is controlled). This happens to be the zone where other sodium channel-modifying mutations (both pyrethroid resistance and heritable disease related mutations) are found in S6 and S5 segments in other homologous domains (I, II, III, IV) (Fig 7). Connected together in this 3-dimensional picture (*41*), these mutations form "a ring" just below the central constriction of the main channel (= chamber). The importance of this "ring" position becomes apparent when one realizes that two of the human hyperkelamic periodic paralysis cases are caused by mutations conferring shifts in amino acids in this ring location as well.

As for the question on the possibility of the *kdr* site being either the binding site for DDT-pyrethroids, or the compensatory counter-balancing site, this new model of Sato et al. (*41*) may be considered to render the support for the former possibility, since the location on the inner wall of the main channel happens to be similar to the one we observed in the case of cyclodiene resistance. However, one interesting observation is that in every case of *kdr*-like pyrethroid resistance mutations involving S6 and S5 segments, the shift of amino acid involves the replacement of a small, and in some cases, polar amino acid with a bulky and more hydrophobic amino acid, in contrast to the case of the cyclodiene resistance. If one relies on a simple assumption that lipophic insecticides readily bind to hydrophobic sites, this observation does not help the former theory. As for the possibility of the *kdr* mutation conferring compensatory counter balancing, at least in one case of *kdr*-like mutation, such a shift in amino acid is likely to act in this manner. This site is a at the lowever end of IIS5 segment, conferring a shift of T to I, the one discovered to be associated with the pyrethroid resistance in diamondback moth (*39*) and headlice (*38*). This site corresponds to the human hyperkelamic periodic paralysis involving T to M shift (T704M). According to Cannon (*43*), this mutation, which is the most frequently observed type of human hyperkelamic periodic paralysis, causes the sodium channel to shift to the state of hyperpolarization as well as acceleration of recovery from inactivation. These functional changes are expected to counterbalance the action of pyrethroids (see 44).

With regard to the S*kdr* type mutations occurring at S4-5 linkers of domain I, II and III (Fig. 7), the most likely explanation is that such mutations induce the shifts in amino acid which will affect the property of the channel inactivation. This appears to be an analogous situation with K-channel where a number of site-directed mutagenesis studies have established the importance in controlling the inactivation process. The observation by Pittendrigh et al. (*30*) that these mutations offer striking synergism to *para* resistance also supports such a conclusion.

Lowering of the Level of Calmodulin Kinase as a Secondary Supporting Mechanism to DDT and Pyrethroid Resistance

Having established that *kdr* resistance in *B. germanica* is clearly caused by the mutation inducing a L to F shift amino acid at a IIS6 segment, we should further ask a question why then these *kdr* resistant cockroaches show cross-resistance to a number of calcium regulating agents, particularly those known to increase intracellular concentrations of calcium. This phenomenon cannot be explained solely on the basis of the single mutation of the sodium channel, since the destabilizing action of DDT on the sodium channel is known to be counteracted by the increasing concentration of Ca^{2+}, but not by decreasing. Therefore, development of cross-resistance to Ca^{2+} increasing agents (meaning that the effect of increasing Ca^{2+} such as the one caused by A23187 would have less stimulatory effects in *kdr* insects) indicates an adjustment for a wrong direction, if one thinks the sodium channel works alone in isolation. There are several possibilities to explain the above cross-resistance, in one departs from the sodium channel. First, DDT or pyrethroids may affect the calcium channel in addition to the sodium channel. In that case it would add some advantages to the altered sodium channel for the *kdr* insects to reduce the cellular sensitivity to increased Ca^{2+} entry due to the forced opening of the calcium channel. Another possibility is that the primary action of these insecticides on the sodium channel induces the secondary effects on the synaptic transmitted releasing process to increase calcium-mediated excitatory synaptic activities. Since we already know that the *kdr* resistance in *B. germanica* is associated with only one mutation at IIS6 of the sodium channel (i.e. no housefly *skdr* type mutation is involved), the above Ca^{2+}-related cross-resistance is not synonymous to the *skdr* mechanism. Rather, it is likely a second resistance mechanism supporting the *kdr* resistance. Indeed Charalambous and Matsumura (*45*) found in *kdr* resistant houseflies that such Ca^{2+}-related cross-resistance pattern is observed in houseflies as well.

It was originally shown by Charalambous and Matsumura (*45*), by using enzymatic analysis that the *kdr* resistant *B. germanica* as well as *M. domestica* show less calcium calmodulin stimulated kinase (=calmodulin kinase of CaM kinase) activities than do the corresponding susceptible counterparts. More recently, we have made more in-depth studies on the same topic, using molecular biological approaches and some of the highlights will be described below.

For the purpose of finding the mechanistic explanation for the cross-resistance of *kdr* insects to the agents known to disrupt Ca^{2+}

homeostasis (i.e. *kdr* insects are resistant to agents which cause an increase or a decrease in intracellular Ca^{2+}), we decided to examine three separate entities in *B. germanica*. They were: calmodulin, Ca-channel and CaM kinase. The reason for this selection was that changes in property of any one of them had the potential to explain the observed phenomenon of the acquired Ca^{2+}-insensitivity among the *kdr* insects. In all cases, we ran a series of polymerase chain reactions (PCR) with designed primer combinations taken from the most conserved region of each gene among many species. As a result, we were successful in obtaining DNA fragments with many homologous regions to known sequences and sequenced them (data not shown). While the sequence data indicated some very interesting properties of those genes and encoded product proteins (*46*), for the purpose of identifying the qualitative difference between *kdr* and the susceptible *B. germanica*, the results were rather disappointing, since we could not find any mutations in any of the sequences we studied. On the other hand, the availability of DNA sequences for these genes provided us with the means to assess the level of their expressions. By using the PCR generated DNA probes, we were able to run a series of Northern blot assays, and found that the most striking interstrain difference was in the quantity of mRNAs encoding CaM kinase II (Fig. 8). The amount of transcripts of CaM kinase II was much lower in both VT and VP strain (both are *kdr* resistant strains) than those found in CSMA or cyclodiene resistant LP strain. This trend was confirmed through Western blotting using an antibody aimed at both N- and C-terminal region of rat CaM kinase. The result was clear in showing that the level of CaM kinase proteins was much higher in two susceptible strains as compared to those found in either VT or VP strain. Therefore, the conclusion we reached by this approach was that the cause for this cross-resistance is likely related to the low expression of CaM kinase (i.e. quantitative difference). Such a conclusion agrees well with the results of CaM kinase enzyme assay made by Charalambous and Matsumura (*45*).

Conclusions

Our efforts to understand the mechanism of resistance developments were tremendously helped by the phenomenal advancement in molecular biological techniques in recent years. Indeed, without those techniques we would have been still in the dark as to the nature of the point mutations which have conferred such specific

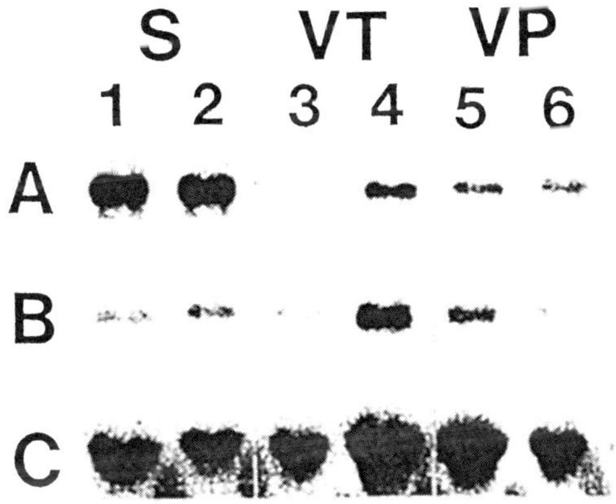

Figure 8. Northern blots indicating the expression of CaM-kinase II gene (A) and L-type calcium channel A1 subunit gene (B) from head and thoraces of Blattela germinaca of susceptible CSMA (S) and kdr resistant (VT and VP) strains. Beta-actin was used as the internal standard. From (46) with permission of Elsevier Science, Inc.

resistance mechanisms to these insects. We feel very fortunate to be in the position to benefit from these advancements and obtain very solid data upon which we can build future research plans. In addition to the availability of modern molecular biological techniques, we are also very fortunate in witnessing the revelation of the functional pictures of ion-channels in action through improved protein isolation, electron microscopy and crystallization which allow electronmicroscopy spin-coupling analyses and x-ray chrystal analyses. While there is only limited information available in this regard, on either the sodium channel or the chloride channel of the GABA receptor, the ACh receptor and the K^+ ion channels studied in detail have revealed a much more dynamic action of the entire subunits of membrane spanning helices than anticipated. These helices go through twisting and turning to create the conformational change needed to open and close the ion-channels in a very orchestrated fashion. The fact that *kdr* mutation in *B. germanica* (*26, 29*) *M. domestica* (*26, 27*) and P. Xylostella (*39*) occurred at the lower 1/3 of the IIS6 segment is very significant. Moreover, in the case of *kdr*-like resistance of H. virescens (*40*) there is yet another mutation at exactly the same site as the kdr mutation to induce a shift from L to H. This confirms the earlier observation that the *kdr*-type shift occurring on one of the S6 or S5 segments requiring a bulkier amino acid as a substitution. In addition the observation that the kdr-like resistance in head lice, *Pediculus capitis* (*38*) that in diamondback moth, *Plutella xylostella* (L) (*39*) and *Heliothis virescens* (*40*) involve mutations also in the lower 1/3 of IIS5 or IS6 segment is very interesting. These are after all the mutations survived in the field despite many odds against them in the real world, unlike the cases of laboratory created mutations. The fact that the IIS6 *kdr* (at 1014, L→F) mutation was confirmed in other insect species (*47-50*) is highly significant from this point of view. These observations render a strong support for the importance of this region in participating in the process of sodium channel inactivation and restoration of the closed state at repolarization. Some of these changes are, at the same time, could compensate for the deleterious effect of DDT and pyrethroids. According to Narahashi (*51*) these insecticides modify both closed (resting) and open state of the sodium channel. Upon depolarization the former converts to the latter first and then much more slowly to the state of inactivation, as compared to the unaffected channel. His studies on pyrethroid-induced prolongation of tail currents in voltage-clamp experiments indicated that the pyrethroid affected channels are kept open longer than normal ones after step repolarization of the membrane. These observations indicated that the mechanism of channel opening and closing is affected by these chemicals in such a way that channel become sluggish in the process of inactivation in responding to the depolarization and the repolarization signals. This phenomenon could be best described as the induced shift in kinetics to resist the channel closure favoring the open state once the channel is activated. The *kdr* modification, if we accept the hypothesis that it is a "compensatory change", is likely to correct the above effect on the channel by shifting the kinetics favoring the closed conformation. Two observations support this view; first the *kdr* modification works better against DDT and type I pyrethroids than against type II pyrethroids which causes much more prolonged tail currents (i.e. keeping it open much longer upon repolarization). If the modification made by type II pyrethroids is much more extensive that the one created by DDT or type I pyrethroids, it would be expected to be more difficult to correct the one caused by type II compounds. Second, the *kdr* modifications offers cross-resistance to even grayanotoxin I and veratrin both of which are known to act on the sodium channel activation process to create a leaky channel as indicated by the loss of Na+ ion selectivity of the channel. That means the *kdr* modification may be able to counteract many types of sodium channel dysfunction leading to higher Na^+ inflow (i.e. leaky channel).

To be sure, we must add that these are largely model building efforts with ample speculations based on the currently available information, and therefore much more work would be needed to confirm, deny, or improve the model proposed herein. This way of model building however, hopefully offers the convenient platform and a fresh viewpoint to chart the next course of research endeavors in search of the molecular basis of the action of these insecticides.

Acknowledgements

This work was supported by grants ES05233 and ES05707 from the National Institute of Environmental Health Sciences and support from the California Agricultural Experiment Station, College of Agricultural and Environmental Sciences, University of California, Davis. We thank Dr. John Marshall Clark for letting us cite the prepublication data on *Pediculus capitis*.

Literature Cited

1. Brown, A.W.A. *Ann. Rev. Entomol.* **1960**, *5*,301-326.
2. Kadous, A.A.; Ghiasuddin, S.M.; Matsumura, F.; Scott, J.G.; Tanaka, K. *Pestic. Biochem. Physiol,* **1983,** *54,*418-540.
3. Ghiasuddin, S.M.; Matsumura, F. *Comp. Biochem. Physiol.* **1982,** *73C,*141-144.
4. Matsumura, F.; Ghisuddin, S.M. *J. Environ. Sci. Health* **1983,***B18,*1-14.
5. Tanaka, K.; Matsumura, F. *In Membrane Receptors and Enzymes as Targets of Insecticide Action*; Clark, J.M. and Matsumura, F., Eds; Plenum Press; New York, NY, **1986**, pp.33-49.
6. Matsumura, F. *In Pesticide Science and Biotechnology. Proceedings of the 6*th *IUPAC Congress of Pesticide Chemistry*; Greenhalgh, R. and Roberts, T.R., Eds; Blackwell Scientific Publications: Oxford, **1987,** pp.151-159.
7. Cole, L.M.; Roush, R.T.; Casida, J.E. *Life Sci.* **1995**, *56*,757-65.
8. ffrench-Constant, R.H.; Pittendrigh, B.; Vaughan, A.; Anthony, N. *Philosophical Transactions of Royal Society of London Series B. Biol. Sci.* **1998**, *353*,1685-1693.
9. Harrison, J.B.; Chen, H.H., Sattelle, E.; Barker, P.J.; Huskisson, N.S., Rauh, J.J.; Bai, D.; Sattelle, D.B. *Cell Tissue Res.* **1996**, *284*, 269-78.
10. Cochran, D.G., *J. Econ. Entomol.* **1994**, *87,*899-903.
11. Ghiasuddin, S.M.; Kadous, A.A.; Matsumura, F. *Comp. Biochem. Physiol.* **1981**, *68C,15-20.*
12. Scott, J.G.; Matsumura, F. *Pestic. Biochem. Physiol.* **1981**, *16*,21-27.
13. Scott, J.G.; Matsumura, F. *Pestic. Biochem. Physiol.* **1983**, *19*, 141-150.
14. Rashatwar, S.; Matsumura, F. *Comp. Biochem. Physiol.* **1985**, *81C*,97-103.
15. Matsumura, F.; Kaku, K.; Enan, E.; Charalambous, P.; Miyazaki, M.; Muralidhara, Inagaki, S. In *Pesticides/Environment: Molecular Biological Approaches*. Mitsui, T.; Matsumura, F., Yamaguchi, I. Eds; Pesticide Science Soc. Japan, **1993**, pp3-15.
16. ffrench-Consultant, R.H.; Rocheleau, T.A.; Steichen, J.C.; Chalmers, A.E. *Nature,* **1993**, *363,*449-451.

17. Thompson, M.; Shotkiski, F.; ffrench-Constant, R.H. *FEBES Leh.* **1993**, *325*, 187-190.
18. Kaku, K.; Matsumura, F. *Comp. Biochem. Physiol.* **1994**, *108C*, 367-376.
19. ffrench-Constant, R.H., Mortlock, D.P., Shaffer, C.D.; MacIntyre, R.J.; Roush, R.T. *Proc. Natl. Acad. Sci. USA* **1991,** *88*, 7209-7213.
20. Matsumura, F.; Hayashi, M. **1966**, *Science 153*, 757-759.
21. Telford, J.N.; Matsumura, F. *J. Econ. Entomol.* **1971**, *64, 230-238.*
22. Matsumura, F.; Tanaka, K. In *Cellular and Molecular Toxicology*. Narahashi, T. Ed; **1984**, pp225-240.
23. Deng, Y.; Palmer, C.J.; Casida, J.E. *Pestic. Biochem. Physiol.* **1991**, *41*, 60-65.
24. Scott, J.G.; Ramaswamy, S.B.; Matsumura, F.; Tanaka, K.J., *Econ. Entomol.* **1986,** *79*, 571-575.
25. Rashatwar, S.; Ishikawa, Y.; Matsumura, F. *Comp. Biochem. Physiol.* **1987**, *88C*, 165-170.
26. Miyazaki, M.; Ohyama, K.; Dunlap, D.Y.; Matsumura, F. *Molec. Gen. Genet.* **1996**, *252*, 61-68.
27. Williamson, M.S.; Martinez-Torres, D.; Hick, C.A.; Devonshire, A.L. *Molec. Gen. Genet.* **1996**, *252*, 51-60.
28. Smith, T.J.; Lee, S.H., Ingles P.J. Knipple, D.C.; Soderlund, D.M. *Ins. Biochem. Molec. Biol.* **1997** *27*, 807-812.
29. Dong, K. *Ins. Biochem. Molec. Biol.* **1997** *27*, 93-100.
30. Pittendrigh, B.; Reenam, R.; ffrench-Constant, R.H.; Ganetzky, B. *Molec. Gen. Genet.*, **1997,** *256,* 602-610.
31. Sato, C.; Matsumoto, G. *Membrane Biol.* **1995** *147*, 45-70.
32. Unwin, N. *Nature* **1995** *373*, 37-43.
33. Perozo, E.; Cortes, D.M.; Cuello, L.G. *Science* **1999,** *285*, 73-78.
34. Gerstein, M.; Chothia, C. *Science* **1999,** *285*, 1682-1683.
35. Chang, C.P.; Plapp, F.W. *Pestic. Biochem. Physiol.* **1983,** *20*, 86-91.
36. Cuello, L.G.; Romero, J.G.; Cortez, D.M.; Perozo, E. *Biochemistry*, **1998,** *37*, 3229-3236.
37. Bendahhou, S.; Cummis, T.R.; Tawil, R.; Waxman, S.G.; Placek, L.J. *J. Neurosci.* **1999**, *19*, 4762-4771.
38. Lee, S.H.; Yoon, K.S.; Williamson, M.S.; Goodson, S.J.; Takano-Lee, M.; Edman, J.D.; Devonshire, A.L.; Clark, J.M.; *Pestic. Biochem. Physiol.* 2000, *66*, 130-43.
39. Schuler, T.H.; Martinez-Torres, D.; Thompson, A.J.; Denholm, I.; Devonshire, A.L.; Duce, I.R.; Williamson, M.S. *Pestic Biochem. Physiol.* **1998,** *59*, 169-182.
40. Park, Y.; Taylor, M.F.J.; Feyereisen, R. *Ins. Mil. Biol.* **1999,** *8,* 161-170

41. Charalambous, P.; Matsumura, F. *Insect Biochem. Mol. Biol.* **1992,** *22,* 721-734.
42. Inagaki, S.; Kaku, K.; Dunlap, D.Y.; Matsumura, F. *Comp. Biochem. Physiol.* **1998,** *C120,*225-233.
43. Guerrero, F.D.; Jamroz, R.C.; Kammlah, D.; Kunz, S.E. *Insect Biochem. Mol. Biol.* **1997,** *27,*745-755.
44. Martinez-Torres, D.; Chandre, F.; Williamson, M.S.; Daniel, F.; Berge, J.B.; Devonshire, A.L.; Gillet, P., Pasteur, N,; Pauron, D. *Insect Mol. Biol.* **1998,** *7,*179-184.
45. Lee, S.H.; Dunn, B.; Clark,J.M.; Soderlund, D.M.; *Pestic. Biochem. Physiol.* **1999,** *63,* 63-75.
46. Smith, T.J.; Lee, S.H.; Ingles, P.J.; Knipple, D.C.; Soderlund, D.M.; *Insect Biochem. Mol. Biol.* **1997,** *27,* 807-812.
47. Narahashi, T. *Neurotoxicology,* **1985,** *6(2),*3-25.

Chapter 4

The Molecular Mechanism of Knockdown Resistance

David M. Soderlund[1], Patricia J. Ingles[1,2], Si Hyeock Lee[1,3], Timothy J. Smith[1], and Douglas C. Knipple[1]

[1]Department of Entomology, New York State Agricultural Experiment Station, Cornell University, Geneva, NY 14456
[2]Current address: Department of Biology, University of Leicester, Leicester LE1 7RH, United Kingdom
[3]Current address: Department of Entomology, University of Massachusetts, Amherst, MA 01003

Knockdown resistance to pyrethroids, first identified and isolated genetically in the house fly, poses a threat to the continued effectiveness of these insecticides against many pest species. Knockdown resistance is linked genetically to a voltage-sensitive sodium channel gene sequence (*Vssc1*) in the house fly that is orthologous to the *para* sodium channel gene of *Drosophila melanogaster*. Sequence analysis of *Vssc1* alleles from susceptible and resistant fly populations has consistently identified two point mutations that are associated with the *kdr* (knockdown resistance) and *super kdr* traits. Functional expression of wildtype and specifically mutated sodium channels in *Xenopus laevis* oocytes has elucidated the effects of resistance-associated mutations on sodium channel function and sensitivity to pyrethroids and has confirmed that these two point mutations are the cause of the *kdr* and *super-kdr* traits.

© 2002 American Chemical Society

Knockdown resistance to DDT and pyrethroids in the house fly (*Musca domestica* L.) was the first insecticide resistance mechanism involving a change in the sensitivity of the target site to be identified (*1*). Genetic linkage analysis located both the knockdown resistance trait (called *kdr*) and a related variant exhibiting greater resistance to some synthetic pyrethroids (called *super-kdr*) to the third chromosome of the house fly (*2-4*), thereby implying that these two traits are conferred by alleles of the same resistance gene. Because pyrethroids are known to exert their insecticidal effects by altering the function of voltage-sensitive sodium channels in nerve membranes (*5,6*), efforts to identify the mechanism underlying knockdown resistance have focused on alterations in the expression or pharmacology of voltage-sensitive sodium channels (*1,5,7*). The most widely-held hypothesis in these studies was that the *kdr* and *super-kdr* traits resulted from mutations in the house fly gene encoding sodium channels that are target sites for pyrethroid action. In this chapter, we review recent research to test this hypothesis by defining the mechanism of knockdown resistance in the house fly at the genetic and molecular level.

Molecular Genetics

Linkage Analysis

The identification of the *para* gene of *Drosophila melanogaster* as a structural gene encoding voltage-sensitive sodium channel α subunits with important physiological roles (*8*) provided a point of entry to explore the genetic relationship between house fly gene sequences orthologous to *para* and the *kdr* and *super-kdr* traits. A polymerase chain reaction (PCR)-based homology probing strategy yielded a short house fly DNA sequence with a high degree of nucleotide sequence similarity to the corresponding *para* sequence (*9*). This PCR-derived fragment was used as a probe to isolate a genomic DNA clone corresponding to a segment of the *para*-orthologous sodium channel α subunit gene of the house fly (*10*).

We identified a restriction fragment length polymorphism (RFLP) marker within the house fly sodium channel gene (*Vssc1*; also called *Msc*) that distinguished between the susceptible and *kdr* genotypes and employed this marker in combination with discriminating dose paralysis assays on individual flies to define the degree of linkage between this gene and the *kdr* trait (*10*). Results of this analysis documented tight genetic linkage between the *kdr* trait and the *Vssc1* gene located on chromosome 3 at a resolution of ~1 map unit (*10*). Using a similar approach, Williamson and colleagues (*11*) showed that the *super-kdr* trait was also tightly linked (within ~1 map unit) to the *Vssc1* gene.

These results provided strong genetic evidence for mutations in a sodium channel structural gene as the cause of knockdown resistance. They also provided experimental evidence for the widely-presumed allelism of the *kdr* and *super-kdr* traits.

Identification of Resistance-Associated Mutations

The documentation of genetic linkage between knockdown resistance and sodium channel gene sequences in the house fly stimulated the complete characterization of the house fly *Vssc1* gene. Independent efforts by two laboratories (*12,13*) resulted in the determination of the complete *Vssc1* cDNA sequence and documented its close relationship to the *para* gene of *D. melanogaster*. Comparison of partial or complete *Vssc1* sequences from 15 house fly strains representing susceptible, *kdr*, and *super-kdr* phenotypes (*13,14*) identified two amino acid sequence polymorphisms that were invariably associated with knockdown resistance (Figure 1). All resistant strains, whether *kdr* or *super-kdr*, contained a mutation of leucine to phenylalanine at amino acid residue 1014 (designated L1014F). In addition, all *super-kdr* strains contained a second mutation, methionine to threonine at residue 918 (M918T). Interestingly, the M918T mutation has not been found as a single polymorphism in the absence of the L1014F mutation in any house fly strain, resistant or susceptible, examined to date.

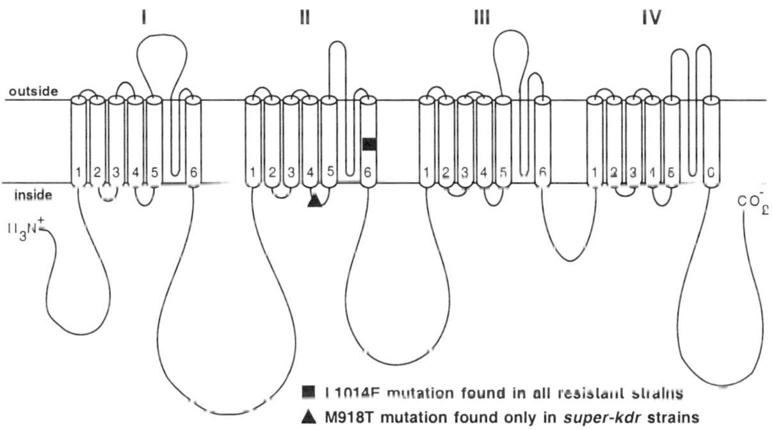

Figure 1. Extended topological model of the sodium channel α subunit protein showing the location of mutations associated with knockdown resistance.

The mutations associated with the *kdr* and *super-kdr* traits imply that these two phenotypes did not arise independently. The involvement of a second-site mutation in the enhanced resistance exhibited by *super-kdr* insects implies that *super-kdr* populations arose from further selection of populations already carrying the *kdr* trait. This interpretation is consistent with the results of RFLP analyses of the *Vssc1* locus of susceptible, *kdr*, and *super-kdr* fly strains, which show extensive RFLP polymorphisms among susceptible strains but strong conservation of RFLP patterns among *kdr* and *super-kdr* populations of diverse geographic origin (*15*). The conservation of RFLP patterns in resistant strains also is consistent with the hypothesis that knockdown resistance had a single genetic origin and has been spread subsequently by dispersal of resistant flies (*15*).

The identification of specific amino acid substitutions associated with the *kdr* and *super-kdr* traits implies that these mutations are the cause of resistance but does not, in itself, provide definitive proof. Moreover, these findings raise several questions that cannot be answered by genetic linkage analysis. The absence of the M918T single mutation among resistant populations suggests that the effect of this mutation is not simply additive with that of the L1014F mutation. Instead, the M918T could conceivably function as an enhancer of the effect of the M918T mutation that has no impact on the pyrethroid sensitivity of house fly sodium channels when present alone. Alternatively, the M918T mutation could confer enhanced resistance at the cost of a functional deficit on house fly sodium channels that is ameliorated in some way by the presence of the L1014F mutation.

Functional Characterization of Resistance-Associated Mutations

Resistance-Associated Mutations in House Fly Sodium Channels

To demonstrate a causal relationship between sodium channel gene mutations and resistant phenotypes, and to understand the functional relationships betweeen the L1014F and M918T mutations, it is necessary to obtain the functional expression of house fly sodium channels and to determine the effects of these amino acid substitutions, inserted singly or in combination by site-directed mutagenesis into the wildtype sequence, on the functional properties and pyrethroid sensitivity of the expressed channel. Expression in oocytes of the frog *Xenopus laevis* (*16*), coupled with electrophysiological assays of the function and pharmacology of the expressed channel (*17*), is the system that has been most commonly employed for structure - function studies of vertebrate sodium channels (*18,19*). Recent studies showed that the *Xenopus* oocyte system also effectively expresses *D. melanogaster* para sodium channels and that both the biophysical properties of the expressed channels closely paralleled those of sodium channels in native neuronal preparations (*20,21*). Moreover, these studies demonstrated that coexpression of the para sodium channel α subunit with a second protein, the product of the *tipE* gene of *D. melanogaster*, greatly enhanced the level of sodium current expression in oocytes (*20,21*).

To take advantage of this experimental approach, we constructed a full-length *Vssc1* cDNA for use as a template for the synthesis of cRNA for expression experiments. Oocytes injected with *Vssc1* cRNA expressed very small voltage-gated sodium currents, but coinjection with *D. melanogaster tipE* cRNA produced robust sodium currents, presumably carried by heteromultimeric channels formed from the Vssc1 and tipE proteins, that were amenable to functional and pharmacological characterization (*22*). These hybrid *Musca/Drosophila* sodium channels, which we term Vssc1/tipE channels, exhibit biophysical properties similar to para/tipE sodium channels expressed in oocytes and sensitivity to blockade by sub-micromolar concentrations of tetrodotoxin (*22,23*). Moreover the modifications of Vssc1/tipE sodium channels by pyrethroids such as cismethrin and [1*R*,*cis*,α*S*]-cypermethrin measured under two-electrode voltage clamp are generally similar to the effects of these and other pyrethroids on sodium currents in voltage-clamped nerve preparations (*23*).

We assessed the impact of the L1014F and M918T mutations on the pyrethroid sensitivity of house fly sodium channels by introducing these polymorphisms into the wildtype *Vssc1* cDNA and comparing the sensitivity of wildtype and specifically mutated Vssc1/tipE sodium channels to cismethrin and [1*R*,*cis*,α*S*]-cypermethrin. The effects of cismethrin on sodium currents recorded under voltage clamp conditions from oocytes expressing either wildtype Vssc1/tipE channels or those carrying the L1014F mutation are shown in Figure 2 (*22*). In assays with wildtype Vssc1/tipE channels, cismethrin produced a sustained sodium current during a 50-msec depolarization and also induced a large, biphasic sodium tail current that flowed at the end of a depolarizing pulse. In comparable assays with channels containing the L1014F mutation, both the amplitudes of both the sustained current during depolarization and the tail current were reduced relative to that of the peak transient sodium current, indicating a reduction in the proportion of cismethrin-modified channels contributing to the observed current. Results of multiple experiments at different cismethrin concentrations showed that channels containing the L1014F mutation were at least 10-fold less sensitive to cismethrin than wildtype channels (*22*). The use-dependent modification of Vssc1/tipE sodium channels by [1*R*,*cis*,α*S*]-cypermethrin and the extremely persistent tail currents caused by this compound (*23*) complicated its use to assess the functional impact of knockdown resistance mutations. Nevertheless, a more limited number of assays with [1*R*,*cis*,α*S*]-cypermethrin corroborated the degree of pyrethroid resistance conferred by the L1014F mutation (*24*).

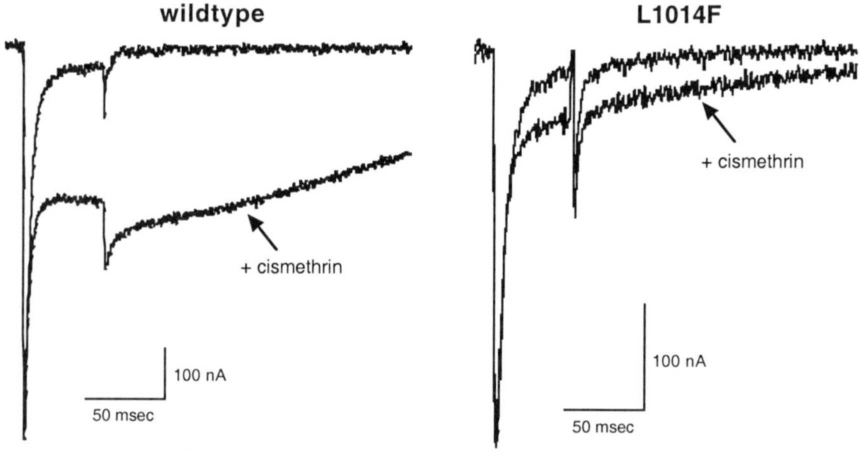

Figure 2. Effects of cismethrin (40 μM) on sodium currents recorded from wildtype Vsscl/tipE sodium channels and from channels containing the L1014F mutation; drawn from data of Smith et al. (22).

We also performed similar experiments with Vsscl sodium channels mutated to contain the M918T/L1014F double mutation associated with the *super-kdr* resistance phenotype. We were unable to detect any modification of sodium currents in oocytes expressing these mutated channels by either cismethrin or [1*R*,*cis*,α*S*]-cypermethrin at the highest concentrations attainable in our assay system (*25*). These results suggest that the two mutations associated with the *super-kdr* phenotype confer a very high level of pyrethroid resistance.

Further insight into the functional significance of the L1014F and M918T/L1014F mutations is available from recent studies by Vais and colleagues (*26*), who inserted these mutations at the corresponding positions of the *D. melanogaster* para sodium channel and determined the sensitivity to deltamethrin of wildtype and specifically mutated para/tipE sodium channels expressed in oocytes. These studies showed that the extent of modification of channels by deltamethrin, like [1*R*,*cis*,α*S*]-cypermethrin, was dependent on repeated channel activation and documented ~20-fold resistance of the L1014F variant to deltamethrin. In contrast to the results obtained with the M918T/L1014F variant of the Vssc1/tipE sodium channel, Vais *et al.* (*26*) were able to detect modification of para/tipE sodium channels containing the M918T/L1014F double mutation by deltamethrin following repeated activation. Under these conditions assay conditions, the double mutation conferred ~100-fold resistance to this pyrethroid. These authors concluded that resistance to deltamethrin in the mutated channels was due to reduced availability of open channels, which exhibit preferential binding of deltamethrin, and enhanced dissociation of deltamethrin from channels in the open state.

The oocyte expression system also permitted studies of the functional and pharmacological impact of the M918T single mutation, which has only been found in fly populations in combination with the L1014F mutation. We found house fly (Vssc1/tipE) sodium channels containing only the M918T mutation to be poorly expressed in oocytes and highly resistant to both cismethrin and [1*R*,*cis*,α*S*]-cypermethrin (*25*). This result suggests that the M918T mutation by itself is sufficient to confer the degree of resistance normally associated with the *super-kdr* phenotype. In view of the high level of pyrethroid resistance conferred by the M918T mutation in our assays, it is surprising that that this single mutation has not been selected in pyrethroid-resistant insect populations. We hypothesize that that the M918T mutation, by itself, confers a functional deficit on insect sodium channels and that the L1014F mutation rescues this deficit sufficiently to make the double mutant viable. A precedent for the correlation of sodium channel functional deficits with mutations at this amino acid sequence position of the sodium channel α subunit is found in studies of paramyotonia congenita, a heritable muscle disease in humans. One form of this disorder is associated with a mutation to threonine at the isoleucine residue of the human skeletal muscle sodium channel α subunit that aligns with Met918 of the Vssc1 sodium channel (*27*).

Resistance-Associated Mutations Inserted into Rat Peripheral Nerve Sodium Channels

To assess the general significance of the mutations associated with the *kdr* and *super-kdr* traits as determinants of pyrethroid sensitivity, we introduced the corresponding mutations into the cloned rat tetrodotoxin-resistant peripheral nerve (SNS) sodium channel α subunit (*28,29*), which is highly sensitive to pyrethroids (*30*), and determined the pyrethroid sensitivity of native and mutated channels expressed in *Xenopus* oocytes (S. H. Lee and D. M. Soderlund, manuscript in preparation). The leucine residue in domain IIS6 that corresponds to the site of the Leu1014 in Vssc1 sodium channels is conserved in all known vertebrate sodium channel α subunit isoforms, but vertebrate sodium channel isoforms contain a conserved isoleucine at sequence positions corresponding to the conserved Met918 of Vssc1 and other insect channels. To test the impact of the *kdr* and *super-kdr* mutations in this molecular context, we examined three specifically mutated SNS cDNAs: the *kdr* mutation at the conserved leucine in domain IIS6 (L879F); the I780T/L879F double mutation that is associated with the *super-kdr* trait; and, the I780T single mutant, which corresponds to the M918T mutation in the Vssc1 sequence.

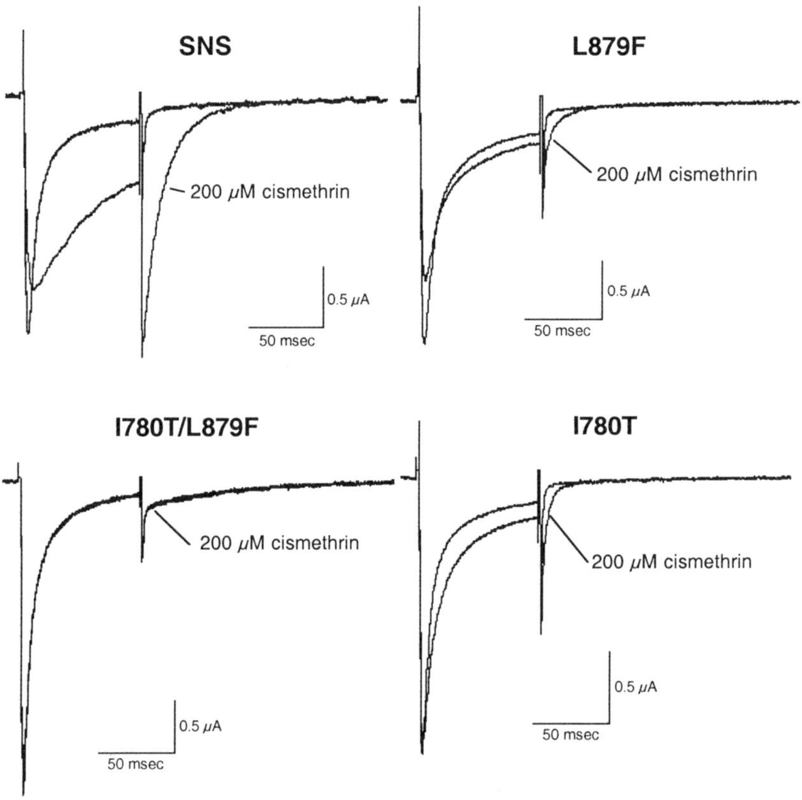

Figure 3. Effects of cismethrin on native (SNS) and specifically mutated rat peripheral nerve sodium channels (S. H. Lee and D. M. Soderlund, unpublished results, 1999).

Figure 3 illustrates the effects of a high concentration of cismethrin on sodium currents recorded from oocytes expressing native or specifically mutated SNS sodium channels. The effects of the L879F mutation associated with the *kdr* trait paralleled the effects this mutation in Vsscl/tipE sodium channels, reducing the threshold sensitivity for cismethrin approximately 10-fold and accelerating the rate of tail current decay. Similarly, the I780T/L879F double mutation, which corresponds to the mutations associated with the *super-kdr* trait, completely abolished the cismethrin sensitivity of SNS channels. This result confirmed that the high level of cismethrin resistance afforded by the double mutation is not limited to the sequence context of insect (Vsscl and para) sodium channel α subunits. Finally, the I780T mutation also significantly reduced the cismethrin sensitivity of SNS channels, thus confirming the functional impact of this single mutation. The results of our studies with mutated SNS channels confirm in an independent molecular context the functional significance of the mutations associated with the *kdr* and *super-kdr* traits. Moreover, these results show that the domains identified by these mutations are determinants of pyrethroid sensitivity in phylogenetically divergent sodium channels.

Conclusions

Structural and functional studies with house fly sodium channels provide convincing evidence that a single point mutation in the house fly *Vsscl* gene is the cause of *kdr* trait in this species. The corresponding mutation also has been found in pyrethroid-resistant strains of several other pest species, (*14,31-37*), making this mutation the single most commonly-encountered cause of knockdown resistance. The practical value of this information is illustrated by the development of a single-insect DNA diagnostic assays capable of detecting this polymorphism for use in the monitoring and management of pyrethroid resistance several insect species (*33,37-39*).

In contrast to the many examples in other species of mutations corresponding to the L1014F mutation in the house fly, a mutation corresponding to the second-site M918T mutation associated with the *super-kdr* trait of the house fly has been found to date only in highly resistant populations of *H. irritans* (*32*). Expression experiments in oocytes show that this mutation, by itself, confers a high level of pyrethroid resistance both in house fly (Vsscl/tipE) and rat (SNS) sodium channels. In light of the significant reduction in pyrethroid sensitivity conferred by the M918T mutation alone, it is likely that this single mutation confers a functional deficit that prevents its selection in populations under pyrethroid pressure. Further research is required to determine the nature of the putative functional deficit conferred by this mutation, its relationship to the alterations in function caused by mutations to threonine at the corresponding position of the human skeletal muscle sodium channel that are considered to cause one form of paramyotonia congenita (*27*), and the means by which the L1014F mutation complements such a deficit to permit selection of the M918T substitution as a second-site mutation in *super-kdr* insects.

The search for sodium channel gene mutations associated with knockdown resistance also has identified numerous novel mutations that were not identified in resistant house fly populations (*33,34,37,40-44*). The diversity of mutations associated with knockdown resistance to DDT and pyrethroids is therefore quite different from the situation encountered with target site resistance to cyclodiene insecticides, which is universally correlated with a point mutation at a single amino acid residue of the *Rdl* subunit of insect γ-aminobutyric acid receptors (*45*). Functional confirmation of reduced sodium channel sensitivity has been obtained for at least one additional resistance associated mutation, the V410M mutation found in some knockdown-resistant populations of *Heliothis virescens* (*46*), but the effects of most of these novel mutations on the functional and pharmacological properties of insect sodium channels remain to be assessed.

The development of resistance in pest populations remains the principal threat to the continued effectiveness of pyrethroid insecticides. Pyrethroid resistance may result from one of several mechanisms, but resistance resulting from reduced target site sensitivity, exemplified by the *kdr* and *super-kdr* traits in the house fly, is particularly problematic because it diminishes the effectiveness of all members of the pyrethroid class (*7*). The well-documented cross resistance of knockdown-resistant insects to DDT also implies that this type of resistance may have already been selected in pest populations as the result of widespread DDT use, thereby leading to the rapid re-selection of resistant alleles under heavy pyrethroid pressure. Continued research to identify and characterize resistance-causing mutations will facilitate both the development of new DNA diagnostic assays for resistance-conferring alleles in pest populations and the identification and wise use of other insecticidal agents that act on insect sodium channels but are unaffected by knockdown resistance mutations.

Acknowledgments

Research from our laboratories summarized in this review was supported in part by grants (94-37302-0408 and 97-35302-4323) from the United States Department of Agriculture NRI/CGP and a grant (R01-ES08967) from the National Institute of Environmental Health Sciences.

References

1. Soderlund, D. M.; Knipple, D. C. *Ann. Ent. Soc. Amer.* **1999**, *92*, 909-915.
2. Milani, R.; Travaglino, A. *Riv. Parassitol* **1957**, *18*, 199-202.
3. Farnham, A. W. *Pestic. Sci.* **1977**, *8*, 631-636.
4. Farnham, A. W.; Murray, A. W. A.; Sawicki, R. M.; Denholm, I.; White, J. C. *Pestic. Sci.* **1987**, *19*, 209-220.
5. Bloomquist, J. R. In *Reviews in Pesticide Toxicology*; Roe, M. and Kuhr, R. J., Ed.; Toxicology Communications: Raleigh, NC, 1993; Vol. 2; pp 181-226.
6. Soderlund, D. M. In *Pyrethrum Flowers: Production, Chemistry, Toxicology, and Uses*; Casida, J. E. and Quistad, G. B., Ed.; Oxford University Press: New York, 1995; pp 217-233.
7. Soderlund, D. M. In *Molecular Mechanisms of Resistance to Agrochemicals*; Sjut, V., Ed.; Springer: Berlin, 1997; pp 21-56.
8. Loughney, K.; Kreber, R.; Ganetzky, B. *Cell* **1989**, *58*, 1143-1154.
9. Knipple, D. C.; Payne, L. L.; Soderlund, D. M. *Arch. Insect Biochem. Physiol.* **1991**, *16*, 45-53.
10. Knipple, D. C.; Doyle, K. E.; Marsella-Herrick, P. A.; Soderlund, D. M. *Proc. Natl. Acad. Sci. USA* **1994**, *91*, 2483-2487.
11. Williamson, M. S.; Denholm, I.; Bell, C. A.; Devonshire, A. L. *Mol. Gen. Genet.* **1993**, *240*, 17-22.
12. Ingles, P. J.; Adams, P. M.; Knipple, D. C.; Soderlund, D. M. *Insect Biochem. Mol. Biol.* **1996**, *26*, 319-326.
13. Williamson, M. S.; Martinez-Torres, D.; Hick, C. A.; Devonshire, A. L. *Mol. Gen. Genet.* **1996**, *252*, 51-60.
14. Miyazaki, M.; Ohyama, K.; Dunlap, D. Y.; Matsumura, F. *Mol. Gen. Genet.* **1996**, *252*, 61-68.
15. Williamson, M. S.; Bell, C. A.; Denholm, I.; Devonshire, A. L. In *Molecular Action and Pharmacology of Insecticides on Ion Channels*; Clark, J. M., Ed.; American Chemical Society: Washington, DC, 1995; pp 86-96.
16. Goldin, A. L. *Meth. Enzymol.* **1992**, *207*, 266-297.
17. Stühmer, W. *Meth. Enzymol.* **1992**, *207*, 319-339.
18. Catterall, W. A. *Physiol. Rev.* **1992**, *72*, S15-S48.
19. Catterall, W. A. *Annu. Rev. Biochem.* **1995**, *64*, 493-531.
20. Feng, G.; Deak, P.; Chopra, M.; Hall, L. M. *Cell* **1995**, *82*, 1001-1011.
21. Warmke, J. W.; Reenan, R. A. G.; Wang, P.; Qian, S.; Arena, J. P.; Wang, J.; Wunderler, D.; Liu, K.; Kaczorowski, G. J.; Van der Ploeg, L. H. T.; Ganetzky, B.; Cohen, C. J. *J. Gen. Physiol.* **1997**, *110*, 119-133.
22. Smith, T. J.; Lee, S. H.; Ingles, P. J.; Knipple, D. C.; Soderlund, D. M. *Insect Biochem. Mol. Biol.* **1997**, *27*, 807-812.

23. Smith, T. J.; Ingles, P. J.; Soderlund, D. M. *Arch. Insect Biochem. Physiol.* **1998**, *38*, 126-136.
24. Soderlund, D. M.; Smith, T. J.; Lee, S. H.; Ingles, P. J.; Knipple, D. C. In *Progress in Neuropharmacology and Neurotoxicology of Pesticides and Drugs*; Beadle, D. J., Ed.; Royal Society of Chemistry: Cambridge, 1999; pp 46-55.
25. Lee, S. H.; Smith, T. J.; Knipple, D. C.; Soderlund, D. M. *Insect Biochem. Mol. Biol.* **1999**, *29*, 185-194.
26. Vais, J.; Williamson, M. S.; Goodson, S. J.; Devonshire, A. L.; Warmke, J. W.; Usherwood, P. N. R.; Cohen, C. J. *J. Gen. Physiol.* **2000**, in press.
27. Plassart, E.; Eymard, B.; Maurs, L.; Hauw, J. J.; Lyon-Caen, O.; Fardeau, M.; Fontaine, B. *J. Neurol. Sci.* **1996**, *142*, 126-133.
28. Akopian, A. N.; Sivilotti, L.; Wood, J. N. *Nature* **1996**, *379*, 257-262.
29. Sangameswaran, L.; Delagado, S. G.; Fish, L. M.; Koch, B. D.; Jakeman, L. B.; Stewart, G. R.; Sze, P.; Hunter, J. C.; Eglen, R. M.; Herman, R. C. *J. Biol. Chem.* **1996**, *271*, 5953-5956.
30. Smith, T. J.; Soderlund, D. M. *NeuroToxicology* **2000**, (submitted).
31. Dong, K. *Insect Biochem. Mol. Biol.* **1997**, *27*, 93-100.
32. Guerrero, F. D.; Jamroz, R. C.; Kammlah, D.; Kunz, S. E. *Insect Biochem. Mol. Biol.* **1997**, *27*, 745-755.
33. Martinez-Torres, D.; Chandre, F.; Williamson, M. S.; Darriet, F.; Bergé, J. B.; Devonshire, A. L.; Guillet, P.; Pasteur, N.; Pauron, D. *Insect Mol. Biol.* **1998**, *7*, 179-184.
34. Schuler, T. H.; Martinez-Torres, D.; Thompson, A. J.; Denholm, I.; Devonshire, A. L.; Duce, I. R.; Williamson, M. S. *Pestic. Biochem. Physiol.* **1998**, *59*, 169-192.
35. Lee, S. H.; Dunn, J. B.; Clark, J. M.; Soderlund, D. M. *Pestic. Biochem. Physiol.* **1999**, *63*, 63-75.
36. Martinez-Torres, D.; Foster, S. P.; Field, L. M.; Devonshire, A. L.; Williamson, M. S. *Insect Mol. Biol.* **1999**, *8*, 339-346.
37. Martinez-Torres, D.; Chevillon, C.; Brun-Barale, A.; Bergé, J. B.; Pasteur, N.; Pauron, D. *Pestic. Sci.* **1999**, *55*, 1012-1020.
38. Guerrero, F. D.; Kunz, S. E.; Kammlah, D. *J. Med. Entomol.* **1998**, *35*, 710-715.
39. Clark, J. M.; Zhang, A.; Dunn, J.; Yoon, K. *Pestic. Sci.* **1999**, *55*, 606-608.
40. Park, Y.; Taylor, M. F. J. *Insect Biochem. Mol. Biol.* **1997**, *27*, 9-13.
41. Park, Y.; Taylor, M. F. J.; Feyereisen, R. *Biochem. Biophys. Res. Commun.* **1997**, *239*, 688-691.
42. Pittendrigh, B.; Reenan, R.; ffrench-Constant, R. H.; Ganetzky, B. *Mol. Gen. Genet.* **1997**, *256*, 602-610.
43. Head, D. J.; McCaffery, A. R.; Callaghan, A. *Insect Mol. Biol.* **1998**, *7*, 191-196.

44. Lee, S. H.; Yoon, K.-S.; Williamson, M. S.; Goodson, S. J.; Takano-Lee, M.; Edman, J. D.; Devonshire, A. L.; Clark, J. M. *Pestic. Biochem. Physiol.* **2000**, *66*, 130-143.
45. ffrench-Constant, R. H. *Insect Biochem. Mol. Biol.* **1994**, *24*, 335-345.
46. Lee, S. H.; Soderlund, D. M. *Insect Biochem. Mol. Biol.* **2000**, (in press).

Chapter 5

A Genomics Perspective on Mutant Aliesterases and Metabolic Resistance to Organophosphates

Charles Claudianos[1], Erica Crone[1,2], Chris Coppin[1,3],]Robyn Russell[1], and John Oakeshott[1,*]

[1]CSIRO Entomology, GPO Box 1700, Canberra, ACT 2601, Australia
Divisions of [2]Botany and Zoology and [3]Biochemistry and Molecular Biology, School of Life Sciences, Australian National University, Canberra, ACT 0200, Australia

There are now five insect taxa for which the molecular mechanisms underlying metabolic resistance to organophosphates have been determined. Essentially only three mechanisms have been found, all involving closely related enzymes in the carboxyl/cholinesterase multigene family. None of them generate an enzyme that is kinetically efficient in degrading a broad range of organophosphates. Paradoxically however several kinetically more efficient enzymes have been found in other eukaryotes and prokaryotes and their homologs are very likely to occur in insects. These homologs may explain several cases of esterase mediated metabolic resistance whose biochemistries are distinct from the five cases so far resolved at a molecular level.

© 2002 American Chemical Society

Changes in carboxylesterase activity have been implicated in many cases of metabolic resistance to organophosphate (OP) insecticides in insects and mites (*1*). However the nature of the change is highly variable, with comparable numbers of cases showing decreases and increases in carboxylesterase activity.

In the early nineties molecular mechanisms were elucidated for two systems in which greatly elevated carboxylesterase activities were associated with resistance (*2, 3*). Although the species, the aphid *Myzus persicae* and culicine mosquitoes, are not closely related, there are close similarities in the mechanisms involved. Both involve massive gene amplifications of up to 300 copies of a particular carboxylesterase gene whose product then accounts for up to 3% of the total protein of the organism. The esterases involved in the two cases are not orthologous but they are members of the same carboxyl/cholinesterase multigene family (*4*). In both cases the enzyme product has almost no OP degradative ability but it binds OP with great avidity and effectively sequesters it.

We have been working to elucidate the molecular bases of OP resistance associated with greatly reduced carboxylesterase activities in two species of higher Diptera, the housefly *Musca domestica* and the sheep blowfly *Lucilia cuprina* (*5, 6, 7*). Their molecular mechanisms also prove to be remarkably similar to each other. Both species use the same amino acid substitutions in orthologous carboxylesterases to create an enzyme with some OP hydrolase activity. The enzyme involved in these species is not orthologous to either the aphid or the mosquito enzymes above, albeit it is in the same multigene family.

Taken together these results suggest that there may be very few molecular options by which carboxylesterases can be used to confer OP resistance. In this paper we consider this proposition in the context of rapidly emerging data on esterase genomics. First however we outline what we have learned about the molecular mechanism of resistance in *M. domestica* and *L. cuprina*.

Mutant Aliesterases In Higher Diptera

Native gel electrophoresis of *M. domestica* and *L. cuprina* reveals polymorphism for orthologous esterase isozymes (termed ALI and E3 respectively) in which variants associated with metabolic resistance to OPs appear as non-staining or only weakly staining when standard artificial substrates like methyl butyrate and naphthyl acetate are used (*8, 9, 10, 11*). The Mutant Aliesterase Hypothesis advanced to explain this proposes that the polymorphism reflects mutation(s) that enable the enzymes to hydrolyse phosphoester bonds like those in OPs at the expense of their ability to hydrolyse carboxylester bonds like those in the artificial substrates (*9,12*).

Newcomb et al. (*5*) confirmed the Mutant Aliesterase Hypothesis by showing that the common non-staining variant of E3 in *L. cuprina* has acquired OP hydrolytic ability as a consequence of a Gly→Asp substitution at residue 137 in its primary sequence. Claudianos et al. (*7*) then showed that the equivalent ALI variant in *M. domestica* has also acquired OP hydrolytic activity as a result of the same substitution at the same residue. Extrapolating from the known three dimensional structure of the related acetylcholinesterase (AChE) enzyme, they showed that the substitution lies in the catalytic centre of the E3/ALI enzyme and indeed is so placed that the alternative Gly and Asp residues could predispose the hydrolysis of carboxylester and phosphoester linkages, respectively.

The parallel between the two species also extends to a second low/non-staining E3/ALI variant associated with a different OP resistance phenotype. The predominant Asp-137 resistance variant hydrolyses the oxon forms of a broad range of OPs but both species also have another less common variant which bestows a somewhat different spectrum of broad resistance among OPs combined with especially high levels of resistance to malathion (*6, 13*). The distinguishing aspect of malathion is that it has two carboxylester bonds in addition to the phosphoester linkages found in all OPs. The malathion resistance variants of E3/ALI in the two species have acquired phosphoester hydrolytic activity while improving the native enzyme's hydrolytic activity against the carboxylester linkages in malathion (albeit with varied losses of activity against the carboxylester linkages of artificial substrates such as methyl butyrate and naphthyl esters). Campbell et al. (*6*) showed that the malathion resistant variant in *L. cuprina* is due to a Trp→Leu substitution at residue 251 in E3 and Claudianos et al. (*7*) then found the same substitution at the same site in ALI. Extrapolation from the AChE structure again localises this site to the catalytic centre, albeit it is unclear in this case how its localisation explains the biochemical differences.

A Ser-251 variant of ALI was also found in *M. domestica*. The strain carrying the variant perished before its OP biochemistry and resistance status could be tested but it may also have been associated with malathion resistance because Ser, like Leu, is a small, uncharged amino acid (*6, 14*).

The levels of malathion resistance conferred by the Leu-251 E3/ALI mutants are around 100 fold and this accords with these enzymes' relatively efficient kinetics for malathion (specificity constants k_{cat}/K_m about 3.4×10^4 $M^{-1}s^{-1}$) (*6, 13, 15*). On the other hand the levels of general OP resistance conferred by both the Asp-137 and the Leu-251 enzymes range from nil up to 30 fold depending on the OP. This concurs with these enzymes' notably inefficient kinetics, with k_{cat}/K_m for one such OP only around 2.3×10^2 $M^{-1}s^{-1}$ and 11 $M^{-1}s^{-1}$, respectively (*6, 13, 15*). These low values for resistance and hydrolysis of general OPs are also notable because as we elaborate below, unrelated enzymes with much more efficient kinetics for OPs have been found in a variety of other organisms.

The Carboxylesterase Multigene Family

The parallels between the resistance mutations in *L. cuprina* and *M. domestica* are all the more remarkable given the large number of related carboxylesterases in which the same (or different) mutations might have arisen. The gene encoding E3/ALI lies in a cluster of very closely related esterase genes called the α-esterase cluster. In *Drosophila melanogaster* where the cluster has been completely characterised there are ten active esterase genes, plus a pseudogene (*16, 17*). The cluster has not been fully ascertained in either *L. cuprina* or *M. domestica* but five putative orthologs have been isolated from the former and eight from the latter (*4, 14, 18*).

Moreover the α-cluster itself will only account for a small minority of the total complement of carboxyl/cholinesterases in each species. At the time of writing the *D. melanogaster* genome project is incomplete but a recent analysis of the *Caenorhabditis elegans* genome project at 90% completion revealed just over 50 carboxyl/cholinesterase sequences spread in singletons or small clusters across just over 30 different sites on its chromosomes (*4*). We know of 18 mapped members of the family in *D. melanogaster* that were discovered independently of that species' genome project. Our preliminary surveys (CCl, EC, CCo, RJR and JGO, unpublished data) of genome project data available at December 1999 (about 90% of the genome) have already yielded another 23 members (Figure 1 and Table I). We predict that there will be close to 50 members of the carboxyl/cholinesterase multigene family in the *D. melanogaster* genome and there seems no reason to expect qualitatively fewer in other invertebrate species.

Many of the carboxyl/cholinesterase sequences in the databases are incomplete but we have screened over 140 complete and non-orthologous sequences from the multigene family from a wide variety of organisms for substitutions at sites aligning to Gly-137 and Trp-251 of E3/ALI (*4*, CCl, RJR and JGO unpublished data). These sequences include over 30 vertebrate genes, 80 from invertebrates, and smaller numbers from lower eukaryotes and bacteria (albeit none from plants).

At the 137 site we find four sequences with Asp, five with Glu, three with Arg, two with Asn, five with Ser and 24 with Ala, with most of the remainder being Gly (about 30 sequences cannot be unambiguously aligned in this region). At the 251 site we find four with Leu, seven with Ser, two with Ileu, four with Ala, four with Gly, three with Thr, three with Tyr and 15 with Phe, with the rest being Trp. No functional data are available for most of the occurrences of Asp and Leu or Ser respectively (and indeed most of all 140 sequences analysed) but several of these occurrences are in, eg vertebrate, species where no selection for metabolism of OP insecticides would have occurred. We also note the occurrences of Glu-137, Ileu-251, Ala-251 and Gly-251, all of which involve amino acids physicochemically similar to Asp-137 and Leu/Ser-251, respectively. We suspect that carboxyl/cholinesterases with Asp or similar residues at site 137 and Leu or similar residues at 251 may often have functions unrelated to OP metabolism.

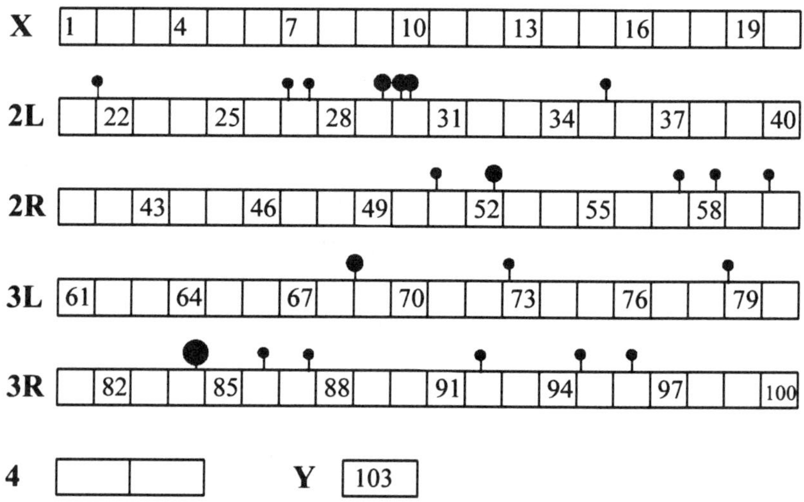

Figure 1. Chromosomal locations of 37 members of the carboxyl/cholinesterase multigene family in D. melanogaster known from prior functional/genetic studies or the D. melanogaster genome project (as at December 1999). Small circles indicate sites of single genes, intermediate circles are for sites of 2-3 genes and the large circle indicates the 11 gene α-cluster (see text). Precise cytogenetic locations, database addresses and names where known are given in Table I.

Table I. 41 carboxyl/cholinesterases from *D. melanogaster*

Chromosome location	*GenBank Accession Number*	*Number of genes*	*Known genes*
22A1	AC005421	1	
27B4-C1	AC004277	1	
27E4-F2	AC009534	1	
29E1-2	AC015133	3	*glutactin*
30B1-10	AC007257	2	
30C1-2	AC013177	2	
35D4	AC003700	1	
51B4-6	AC014449	1	
52E1-8	AC004248	2	
57F1-58A2	AC009912	1	*rel-JHE*
58E1-F2	AC008098	1	*αE1-o*
60B2-10	AC004642	1	
69A1	AC014963	2	*Est6, Est7*
73B5-E2	AC006933	1	*neurotactin*
79A	AC017581	1	
84D9	AC015272	10+1	*αE1-αE10, αE4-Ψ*
86D1-E2	AC006491	1	
87E1-6	AC007754	1	*AChE*
92D1-4	AC008364	1	
95A1-5	AC008236	1	
96D1-E4	AC018022	1	
?	AC018259	1	
?	AC020345	2	
?	AC020118	1	

See Reference 4 for original citations for many of the known genes. *glutactin* and *neurotactin* encode catalytically inactive molecules with signal transduction functions. The genome project reveals that *glutactin* lies in a cluster of three closely related genes at least two of which would encode catalytically inactive proteins. *rel-JHE* encodes a protein with strong sequence similarity to lepidopteran juvenile hormone esterase and may correspond to the *cricklet* gene (*19*). In terms of sequence affinities *αE1-o* belongs to the α-esterase cluster but lies at a remote cytological location (*20*). *αE4-Ψ* is an inactive pseudogene within the α-esterase cluster (*21*). Four members of the family shown with ? have been identified by the genome project but not yet assigned to contigs of known cytogenetic location.

Importantly also there are functional data for a small number of sequences with informative amino acids at 137 or 251. Notable here are some synthetic mutant alleles of human butyrylcholinesterase (BuChE). Knowledge of the tertiary structure of BuChE and, latterly, of the E3/ALI mutations, has been used in attempts to design BuChE variants that might degrade certain OPs that could be used as chemical warfare agents (*4, 22, 23, 24*). Intriguingly an Asp substitution at the residue equivalent to Gly-137 does not confer activity against nerve gas or insecticidal OPs but a His substitution at this site does, albeit again with poor kinetics. Other substitutions at this site or nearby sites in the so-called oxyanion hole region do not. A His at this site was not found in any of the 140 naturally occurring carboxyl/cholinesterase sequences we surveyed above. These data reinforce the E3/ALI evidence that the scope for carboxyl/cholinesterases to acquire OP hydrolytic activity is very restricted.

The same picture emerges from work with EST23, the *D. melanogaster* ortholog of E3/ALI. Although OP resistance does occur in field populations of *D. melanogaster* and can also be selected for after mutagenesis of laboratory populations (*25, 26*) it does not map to the *Est23*/α-esterase cluster. Moreover, preliminary data suggest that a synthetic Asp-137 mutant of EST23 does not have OP hydrolase activity either (W. Odgers, R. Heidari, RJR and JGO unpublished data). Clearly there are very tight structural constraints that determine whether a Gly137→Asp mutation will confer OP hydrolase activity on a carboxyl/cholinesterase.

Less functional information is available around the Trp251→Leu substitution. Notably, Zhu et al. (*27*) have recently recovered a Trp→Gly substitution at the equivalent site in a carboxylesterase from a strain of the parasitoid wasp, *Anisopteromalus calandrae*, that was resistant to malathion. Gly is another small, uncharged amino acid like Leu and Ser. The carboxylesterase in question has several characteristic α-esterase sequence motifs, although the species' distance from the higher Diptera prevents any assignment of orthology with particular members of the higher dipteran cluster. This might suggest that the replacement of a bulky hydrophobic Trp with a smaller residue like Gly, Ser or Leu at this site may enhance malathion hydrolysis in at least a couple of α-esterases.

Finally we also note some of our own preliminary data for the coleopteran *Tribolium castaneum* that has been heavily exposed to OP insecticides. We have used degenerate α-esterase primers to PCR amplify four related carboxylesterase sequences from the malathion resistant GA-1 strain of this species (Genbank Accession Numbers AF260820-AF260823). One of these sequences (AF260820) has both Asp-137 and Leu-251, albeit we have not yet tested whether it has a casual role in hydrolysis and resistance (*14*, CC, B. Campbell, RJR and JGO unpublished data).

Other Esterase Options

At least three distinct OP hydrolysing enzymes have been isolated from bacteria with a history of exposure to OPs (*29*). The two for which sequence data are available are unrelated to each other or to the carboxyl/cholinesterases. One, a prolidase, normally functions in hydrolysis of X-Pro dipeptides. Its activity for insecticidal OPs is reported as modest although we have not seen it reported in terms of k_{cat}/K_m specificity constants (*29*). The other, OP degrading enzyme, OPD, is much more thoroughly characterised and has impressive kinetics for a range of OPs; for example k_{cat}/K_m for paraoxon is about 10^8 $M^{-1}sec^{-1}$, which approaches the diffusion limited maximum (*30, 31, 32*). Its reaction mechanism is not well understood but clearly differs from that of the carboxyl/cholinesterases and directly or indirectly involves metal ions (preferably Zn^{++}). A homolog of OPD has been identified in vertebrate genomes, although its function(s) in vertebrates is not known, and it may well exist in insects as well (*33*). Transgenic *D. melanogaster* expressing bacterial OPD show a level of protection against OP poisoning (*34, 35*), as do heliothis caterpillars infected with a recombinant baculovirus expressing the enzyme (*36*).

Work with mammals has also yielded another quite distinct OP hydrolysing enzyme, termed paraoxonase, or PON1. It is again a metalloenzyme (preferably Ca^{++} in this case), MW around 45KDa, which is associated with low density lipoproteins in plasma and normally involved in metabolism of oxidised lipid compounds (*37, 38*). Its OP hydrolytic activity depends on histidine residues and yields good kinetics (k_{cat}/K_m around 10^6 $M^{-1}s^{-1}$ for paraoxon) (*39, 40*). "Knock-out" mice lacking PON1 show unusual sensitivity to OP poisoning (*41*). The gene encoding PON1 is part of a triplication but the activities of the PON2 and PON3 enzymes for OPs are not reported as yet (*42*). A PON1 homolog has been reported from *C. elegans* and probably exists in insects as well (*43*).

There is also evidence of other, so-called diisopropyl fluorophosphatase (DFP'ase) enzymes in a wide range of vertebrates, invertebrates and microorganisms (*44, 45, 46*). These enzymes are notably diverse in many of their biochemical properties but are all characterised by their hydrolytic activity against OP chemical warfare agents. Limited sequence data suggest they are unrelated to all the other OP hydrolytic enzymes and the carboxyl/cholinesterases above. Although OP nerve agents have very similar structures to many insecticidal OPs, the DFP'ase enzymes are generally described as having negligible or only limited activity against insecticidal OPs. In fact, however, their k_{cat}/K_m data still compare with those for the mutant carboxylesterases in the OP resistant dipterans we have studied.

Conclusions

Only a handful of cases of esterase-based metabolic resistance to OPs in insects have been characterised at a molecular level to date but already we are seeing the recurrence of perhaps just three molecular mechanisms: over-expression of a sequestering enzyme by gene amplification, and generation of a weak oxon OP hydrolytic capability in a detoxifying esterase with (Trp-251→small residue), or without (Gly137→Asp), the generation of high malathion carboxylesterase activity. All three cases involve members, or close relatives of the α-esterase cluster within the carboxyl/cholinesterase multigene family.

However we have also seen that several other completely unrelated esterases can give kinetically much more efficient OP hydrolysis in a range of other organisms and that homologs of these latter enzymes almost certainly also exist in insects. Why then have the cases of OP resistance elucidated thus far involved so few options? Will the same options recur in future analyses of other cases or will some of the other possible mechanisms then be recovered?

There are some reasons to expect at least some recurrence of the mechanisms identified to date. Enzymes encoded by the α-esterase cluster generally seem to be implicated in detoxification of xenobiotics (*4, 20*), so the timing and location of their expression and relatively high abundance may predispose to an effective resistance function. Also, whilst their kinetics are poor in terms of k_{cat}/K_m, this is mainly due to their poor k_{cat}. K_m is quite respectable, around μM values for many OPs, this may be quite important for relatively abundant enzymes to protect the organism effectively against OPs. There is also some evidence that the expression of α-esterases is integrated with other complementary detoxification systems; in *M. domestica* for example there appears to be a genetic association between OP resistant ALI and upregulated cytochrome P450s and glutathione-S-transferases (*14, 47*). In addition, it may be that the size of the α-cluster generates at least some redundancy of function, allowing old functions still to be performed while individual members are recruited to new functions like OP resistance.

On the other hand there are also some good reasons to expect that further work will implicate some new mechanisms which may or may not be based on insect homologs of kinetically apparently superior systems like OPD or PON1. The amplified esterases and mutant aliesterases are certainly not ideal evolutionary solutions. As well as the generally poor detoxification rates, there can also be significant metabolic costs in some cases (*1, 26*). Significantly, very few other cases of esterase based metabolic resistance to OPs have biochemical phenotypes that match those of the amplified aphid and mosquito esterases or the mutant aliesterases of the higher Diptera.

Finally we note that selection for OP resistance has only been operative for at most about fifty years. Therefore the mechanistic options that have been exploited to date have been those that are easily accessible by simple mutations from pre-existing genotypes. In some cases the mutations may even have predated OP insecticides. We have seen that Asp-137 and Leu/Ser/Gly-251 are not uncommon across the carboxyl/cholinesterase multigene family and may have existed in E3/ALI at low frequencies at least before OP insecticides were used. However, if the selection pressure is sustained for some period into the future we might expect less immediately accessible but physiologically more efficient mechanisms to arise. They may involve carboxylesterases or they may not.

Acknowledgements

We thank Narelle Dryden for her help in preparing this paper. We also acknowledge many other current and previous members of our group, in particular Richard Newcomb, Peter Campbell and Charles Robin whose outstanding efforts laid the foundations for the ideas herein. We are grateful to the Australian Cotton, Horticulture, Grains and Rural Industries Research and Development Corporations and to Orica Ltd who all provided financial support for the work reported here.

References

1. Devonshire, A.L.; Field, L.M.; Foster, S.P.; Moores, G.D.; Williamson, M.S.; Blackman, R.L. *Phil. Trans. Roy. Soc. B.* **1998**, 353, 1677-1684.
2. Mutero, A.; Pralavoria, M.; Bride, J-M.; Fournier, D. *Proc. Natl. Acad. Sci. USA*. **1994**, 91, 5922-5926.
3. Field, L.M.; Devonshire, A.L. *Biochem. J.* **1998**, 330, 169-173.
4. Oakeshott, J.G.; Claudianos, C.; Russell, R.J.; Robin, G.C. *BioEssays*. **1999**, 21, 1031-1042.
5. Newcomb, R.D.; Campbell, P.M.; Ollis, D.L.; Cheah, E.; Russell, R.J.; Oakeshott, J.G. *Proc. Natl. Acad. Sci. USA*. **1997**, 94, 7464-7468.
6. Campbell, P.M.; Newcomb, R.D.; Russell, R.J.; Oakeshott, J.G. *Insect Biochem. Mol. Biol.* **1998**, 28, 139-150.
7. Claudianos, C.; Russell, R.J.; Oakeshott, J.G. *Insect Biochem. Mol. Biol.* **1999**, 94, 7464-7468.
8. Ogita, Z.; Kasai, T. *Japan J. Genet.* **1965**, 40, 173-184.
9. Hughes, P.B.; Raftos, D.A. *Bull. Ent. Res.* **1985**, 75, 535-544.
10. Russell, R.J.; Dumancic, M.M.; Foster, G.G.; Weller, G.L.; Healy, M.J.; Oakeshott, J.G. In, *Ecological and Evolutionary Genetics of Drosophila*. Barker, J.S.F.; Starmer, W.T.; MacIntyre, R.J., Eds.; Plenum Press: New York, 1990; pp.293-314.

11. Campbell, P.M.; Trott, J.F.; Claudianos, C.; Smyth, K-A.; Russell, R.J.; Oakeshott, J.G. *Biochem. Genet.* **1997**, 35 , 17-40.
12. Campbell, P.M.; Yen, J.L.; Masoumi, A.; Russell, R.J.; Batterham, P.; McKenzie, J.A.; Oakeshott, J.G. *J. Econ. Entomol.* **1998**, 91, 367-375.
13. Oppenoorth, F.J.; van Asperen, K. *Science.* **1960**, 132, 298-299.
14. Claudianos C. PhD thesis. The Australian National University, Canberra, Australia, 1999.
15. Smyth, K-A.; Boyce, T.M.; Russell, R.J.; Oakeshott, J.G. *Heredity.* **2000**, (in press).
16. Russell, R.J.; Robin, G.C.; Kostakos, P.; Newcomb, R.D.; Boyce, T.M.; Medveczky, K.M.; Oakeshott, J.G. *Insect. Biochem. Mol. Biol.* **1996**, 26, 235-247.
17. Robin, G.C.; Russell, R.J.; Medveczky, K.M.; Oakeshott, J.G. *J. Mol. Evol.* **1996**, 43, 241-252.
18. Newcomb, R.D. PhD thesis. The Australian National University. Canberra, Australia, 1996.
19. Shirras, A.D.; Bownes, M. *Proc.Natl.Acad.Sci. USA.* **1989**, 86, 4559-4563.
20. Robin, G.C.; Claudianos, C.; Russell, R.J.; Oakeshott, J.G. (*unpublished*).
21. Robin, G.C.; Russell, R.J.; Cutler; D.J. Oakeshott, J.G. *Mol. Biol. Evol.* **2000**, (*in press*).
22. Lockridge, O.; Blong, R.M.; Masson, P.; Froment, M.T.; Millard, C.B.; Broomfield, C.A. *Biochemistry.* **1997**, 36, 786-795.
23. Broomfield, C.; Mills, K.V.; Meier, B.M.; Lockridge, O.; Millard, C.B. In, *Structure and Function of Cholinesterases and Related Proteins.* Doctor, B.P.; Taylor, P.; Quinn, D.M.; Rotundo, R.L.; Gentry, M.K. Eds.; Plenum Press: New York, 1998, pp223-226.
24. Oakeshott, J.G.; Claudianos, C.; Campbell, P.M.; Robin, G.C.; Newcomb, R.A.; Odgers, W.A.; Harcourt, R.L.; Russell, R.J. In, *Structure and Function of Cholinesterases and Related Proteins.* Doctor, B.P.; Taylor, P.; Quinn, D.M.; Rotundo, R.L.; Gentry, M.K. Eds.; Plenum Press: New York, 1998, pp 497-502.
25. Oakeshott, J.G.; van Papenrecht, E.A.; Boyce, T.M.; Healy, M.J.; Russell, R.J. *Genetica.* **1993**, 90, 239-268.
26. Batterham, P.; Davies, A.G.; Game, A.Y.; McKenzie, J.A. *BioEssays.* **1996**, 18, 841-845.
27. Zhu, Y.C.; Dowdy, A.K.; Baker, J.E. *Insect Biochem. Mol. Biol.* **1999**, 29, 417-425.
28. Mulbry, W.W.; Karns, J.S. Purification and characterization of three parathion hydrolases from gram-negative bacterial strains. *Appl. Environ. Microbiol.* **1989**, 55, 289-293.
29. Cheng, T-C.; DeFrank, J.J.; Rastogi, V.K. *Chem. Biol. Interact.* **1999**, 119-120, 455-462.
30. Benning, M.M.; Kuo, J.M.; Raushel, F.M.; Holden, M. *Biochemistry.* **1994**, 33, 15001-15007.

31. Lai, K.; Stolowich N.J.; Wild, J.R. *Arch. Biochem. Biophys.* **1995**, 318, 59-64.
32. Di Sioudi, B.D.; Miller, C.E.; Lai, K.; Grimsley, J.K.; Wild, J.R. *Chem. Biol Interact.* **1999**, 119-120, 211-223.
33. Davies, J.A.; Buchman, V.L., Krylova, O.; Ninkina NN. *FEBS Lett.* **1997**, 410, 378-382.
34. Phillips, J.P.; Xin, J.H.; Kirby, K.; Milne, C.P. Jr.; Krell, P.; Wild.; J.R. *Proc. Natl. Acad. Sci. USA.* **1990**, 87, 8155-8159.
35. Benedict, M.Q.; Scott, J.A.; Cockburn, A.F. *Insect Mol. Biol.* **1994**, 3, 247-252.
36. Dumas, D.P.; Wild, J.R.; Raushel, F.M. *Experientia.* **1990**, 46, 729-731.
37. Gan, K.N.; Smolen, A.; Eckerson, H.W.; La Du, B.N. *Drug Metab. Dispos.* **1991**, 19, 100-106.
38. Sorenson, R.C.; Primo-Parmo, S.L.; Kuo, C.L.; Adkins, S.; Lockridge, O.; La Du, B.N. *Proc. Natl. Acad. Sci. USA.* **1995**, 92, 7187-7191.
39. Doorn, J.A.; Sorenson, R.C.; Billecke, S.S.; Hsu, C.; La Du, B.N. *Chem. Biol. Interact.* **1999**, 119-120, 235-241.
40. Hong, S.B.; Raushel, FM. *Chem. Biol. Interact.* **1999**, 119-120, 225-234.
41. Shih, D.M.; Gu, L.; Xia, Y.R.; Navab, M.; Li, W.F.; Hama, S.; Castellani, L.W.; Furlong, C.E.; Costa, L.G.; Fogelman, A.M.; Lusis, A.J. *Nature.* **1998**, 394, 284-287.
42. Primo-Parmo, SL.; Sorenson, R.C.; Teiber, J.; La Du B.N. *Genomics.* **1996**, 33, 498-507.
43. La Du, B.N.; Aviram, M.; Billecke, S.; Navab, M.; Primo-Parmo, S.; Sorenson, R.C.; Standiford, T.J. *Chem. Biol Interact.* **1999**, 119-120, 379-388.
44. Wang, Q.; Sun, M.; Zhang, H.; Huang, C. *J. Biochem. Mol. Toxicol.* **1998**, 12, 213-217.
45. Hoskin, F.C.G.; Walker, J.E.; Mello, C.M. *Chem. Biol. Interact.* **1999**, 119-120, 399-404.
46. Billecke, S.S.; Primo-Parmo, S.L.; Dunlop, C.S.; Doorn, J.A.; La Du, B.N.; Broomfield, C.A. *Chem. Biol. Interact.* **1999**, 119-120, 251-256.
47. Plapp Jr, F.W.; Carino, F.A.; Wei, V.K. *Arch. Insect. Biochem. Physiol.* **1998**, 37, 73-79.

ARTHUR LAKES LIBRARY
COLORADO SCHOOL OF MINES
GOLDEN, CO 80401

Chapter 6

Molecular Approaches to Insect Resistance Management

J. Marshall Clark[1], Si Hyeock Lee[1,3], Hyo-Jeong Kim[1], Kyong Sup Yoon[1], and Aiguo Zhang[2]

[1]Department of Entomology, University of Massachusetts, Amherst, MA 01003
[2]Bio Rad Laboratories, Hercules, CA 94547
[3]Current address: Department of Insect Molecular Biology, Seoul National University, Seoul, Korea

Three DNA-based genotyping techniques, bi-directional PCR amplification of specific allele (bi-PASA), single stranded conformational polymorphism (SSCP) and minisequencing, have been developed and compared for the detection of the S291G (insensitive acetylcholinesterase) and L1014F (insensitive sodium channel) mutations associated with azinphosmethyl and permethrin resistance, respectively, in the Colorado potato beetle (CPB). Extraction of genomic DNA from individual neonates that were hatched from previously collected egg masses is the most efficient and reliable means to obtain suitable templates in terms of convenience, economy, speed, and DNA quality. bi-PASA, employing two allele-specific primers, appears to be the most efficient and rapid genotyping method for the simultaneous detection of both resistant/susceptible homozygous (SS, RR) and heterozygous (SR) alleles. Its resolution, however, is strongly dependent on

© 2002 American Chemical Society

the quality of template genomic DNA. SSCP also allows unambiguous genotyping, including the detection of heterozygous alleles, is less dependent on template DNA quality, however, it requires a longer processing time. Minisequencing is amenable to a 96-well microtiter plate format for the processing of a large number of samples and allows direct detection of resistant/susceptible homozygous alleles but is not as efficient as the PASA and SSCP in detecting heterozygous alleles. In considering the advantages and disadvantages of each technique, DNA-based genotyping is best employed in combinations, with the bi-PASA as the primary method and the SSCP and minisequencing as the secondary validating methods. These methods are rugged, rapid, cost-effective and capable of resolving SS, RR and SR individuals. The availability of such DNA-based genotyping techniques, using neonate genomic DNA as templates, will enable the precise monitoring of the resistant and susceptible allele frequencies, including those of heterozygote individuals, in field populations of CPB. Additionally, an immunoassay based on antibodies raised against esterases associated with resistance has been developed and is considered to be more efficient than a bioassay or enzyme assay because of its higher sensitivity and specificity.

Resistance to azinphosmethyl (AZ-R strain) (1) and carbofuran (BERTS-R strain) (2) in the Colorado potato beetle (CPB) is autosomal, essentially monofactorial and due to a major target site insensitivity associated with an altered acetylcholinesterase (AChE) (3). AChE insensitivity and resistance to these insecticides is associated with a point mutation (A to G, nt location 980) that results in a serine to glycine amino acid change (S291G) in the AZ-R *AChE* gene (4,5). The predicted secondary structure indicates a helix deformation at both the catalytic and peripheral anionic sites due to the S291G mutation and is supported by our detailed biochemical and pharmacological data (6).

Resistance in the permethrin-resistant (PE-R) strain of CPB is sex-linked (1) and due, in part, to a target site insensitivity associated with the nervous system, which is similar, if not identical, to the knockdown resistance (*kdr*) originally identified in house flies (7). We have identified a C to T mutation in the S6 transmembrane segment of domain II (IIS6) of the voltage-sensitive sodium channel α-subunit gene of the PE-R strain (*LdVssc1,* CPB ortholog of the house fly *Vssc1* gene) (8). This mutation results in an amino acid change from a leucine to a phenylalanine (L1014F).

Regardless of what we name them (insecticides, insect growth regulators, agrochemicals, pest control agents, biocides, genetically-modified organisms, etc.), chemicals, in some fashion, will remain the primary method of efficacious insect pest control well into our foreseeable future even under the strengthened EPA/FQPA regulations and increased public pressure. Widespread selection of insect pest populations by agents that kill them, regardless of how, will result in resistance and resistance management in conjunction with IPM will be the principle tools used to suppress or slow this inevitable process. Without efficient, rugged and cheap diagnostic procedures, which can separate susceptible (SS) and resistant (RR) genotypes (including resistant heterozygotes, SR), the goal of effective and economically-feasible resistance management remains largely an unproven computer-generated theory and unattainable for most insect pests.

We have evaluated three separate DNA-based genotyping protocols, bi-directional PCR amplification of specific allele (bi-PASA), single stranded conformational polymorphism (SSCP), and minisequencing, for their ability to identify and to genotype individual beetles that possess or do not possess the S291G and L1014F mutations.

Permethrin resistance in the PE-R strain is also partially due to an elevated hymolymph level of permethrin carboxylesterases (CbEs), which functions in a hydrolytic and sequestration manner (9-11). The permethrin CbEs are inhibited by permethrin and DDT in a manner most similar to a mixed – noncompetitive type and a noncompetitive type of inhibition, respectively, and analysis of these kinetic differences indicate the presence of hydrophobic catalytic site(s) as well as hydrophobic non-catalytic site(s) that are available for the binding of esteratic (e.g., permethrin) and non-esteratic (e.g., DDT) hydrophobic insecticides, respectively (11). We have generated polyclonal antibodies against the three denatured forms of the permethrin CbE that are most responsible for permethrin detoxification and resistance, and used them in an antibody capture immunoassay technique as a possible monitoring tool for the detection of permethrin resistance in field populations of CPB.

We present the advantages and disadvantages of each protocol and summarize with our most efficient strategy to date for monitoring field populations of CPB for resistance management.

Molecular Techniques

DNA Diagnostic Assays

Various stages of CPB, including egg and larval stages, were homogenized using disposable 1.5 ml-tubes and pestles and genomic DNA extracted with DNAzol or DNAzol-BD (Molecular Research Center Inc., Cincinnati, OH) according to the manufacturer's instruction.

In the initial development of DNA-based genotyping protocols, automated DNA sequencing was performed on the mutation-containing regions of several genomic DNA samples isolated from individual beetles for each mutation to validate the genotypes (Automated DNA Sequencing Facility, University of Massachusetts-Amherst). The sequence-verified genomic DNA samples for homozygous susceptible (SS) or resistant (RR) alleles were used as standard templates in the development of DNA-based genotyping protocols. An artificial heterozygous DNA template (SR) was prepared by mixing the homozygous susceptible and resistant allele DNA samples in equal molar ratios and used for PCR.

bi-PASA

The basic methods for bi-PASA were adapted from Williamson *et al.* (12) and Liu *et al.* (13). As shown in the schematic diagram (Fig. 1), bi-PASA reaction contains four separate primers. The two inner primers are allele-specific at the mutation site and the two outer primers are allele-nonspecific. In bi-PASA, one allele is amplified in one direction whereas the other allele is amplified in opposite direction, resulting in two allele-specific fragments that differ in size. All four primers were designed to have similar Tm values (60.7~62.0 °C for S291G bi-PASA; 53.6~54.1 °C for L1014F bi-PASA). The primer sequences and thermal cycler programs employed in bi-PASA are shown in Table 1. Approximately 50~300 ng genomic DNA template was used in a typical 15~20 μl bi-PASA reaction using Taq polymerase (Promega, Madison, WI). Following PCR, amplified DNA fragments were separated by size using 2% Metaphor agarose (FMC Bioproducts, Rockland, ME) gel electrophoresis and visualized by ethidium bromide staining.

Table I. Primer sequences and thermal cycler programs used in the bi-PASA reactions for the detection of the *AChE* S291G and *LdVssc1* L1014F mutations associated with resistance

	AChE S291G mutation	*Ldvssc1* L1014F mutation
Primers		
Resistant allele-specific forward primer	5'CGTGGAGCTACATGTCCG	5'CACAGTTGTTATTGGCAATT
Susceptible allele-specific reverse primer	5'TTGTTCTGCGCGTTCACT	5'GTGGTAATATACTCACAACAAG
Allele-nonspecific forward primer	5'GCTATACGTTGGATCAAAGACA	5'GAATCAATGTGGGACTGT
Allele-nonspecific reverse primer	5'ACTGCTCTCATACAGTCCATCA	5'AACTCGAGACAACATTGATA
Thermal cycler program	(94 °C/ 1 min) × 1 → (94 °C/ 30 sec + 62 °C/ 30 sec + 72°C/ 1 min) × 35	(94 °C/ 1 min) × 1 → (94 °C/ 30 sec + 58 °C/ 30 sec + 72 °C/ 1 min) × 2 → (94 °C/ 30 sec + 57 °C/ 30 sec + 72°C/ 1 min) × 16 → (94 °C/ 30 sec + 56 °C/ 30 sec + 72°C/ 1 min) × 17

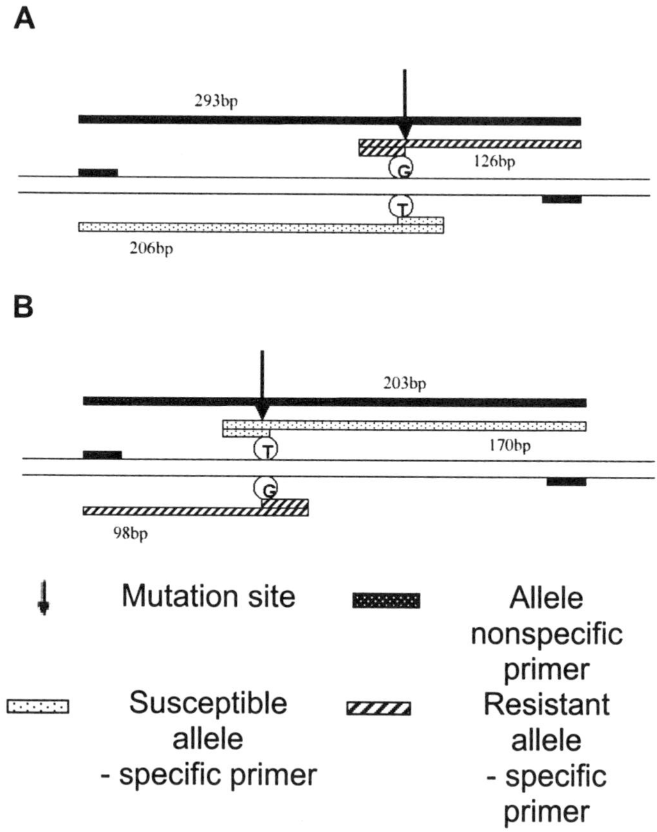

Figure 1. Schematic diagram of bi-PASA reactions for the detection of the *AChE* S291G (A) and *LdVssc1* L1014F (B) mutations.

SSCP

SSCP analysis was based on the methods of Coustau and ffrench-Constant (14) and Hongyo *et al.* (15). Genomic DNA fragments containing the sites of the *AChE* S291G (247 bp) and *LdVssc1* L1014F (349 bp) mutations, respectively, were amplified by PCR using sequence-specific primers:

for *AChE*;
5'GCTATACGTTGGATCAAAGACA *vs.* GAGAAACGTTGCAGCCACAG
for *LdVssc1;*
5'CGAGTACTCTGTGGAGAATGGA *vs.* 5'CCGACACTAGTGATCACCCTTT

The locations of the mutation sites were approximately in the center of respective PCR product. Amplified PCR products were denatured and single stranded DNA separated by polyacrylamide or MDE (FMC Bioproducts) gel electrophoresis at room temperature (for S291G SSCP) or at 4 °C (for L1014F SSCP) as described previously (16). Following electrophoresis, DNA bands were visualized by silver staining (17).

Minisequencing

The basic procedures for solid phase minisequencing were based on those described originally by Syvanen *et al.* (18). The minisequencing reaction for the detection of *AChE* S291G mutation has been described previously (16).

The overall minisequencing procedures for the detection of the *LdVssc1* L1014F mutation were basically identical to those of the S291G minisequencing. A 349-bp biotinylated genomic DNA fragment that contains the L1014F mutation site was amplified with the same set of gene-specific primers used in SSCP except for the 5' biotinylated forward primer (Fig. 2). Antisense detection primer (5'GGTAATATACTCACAACAA) was employed for annealing and digoxigenin-dATP and digoxigenin-dUTP were used for extension reactions (Fig. 2).

Antibody Capture Immunoassay

Polyclonal antibodies were generated against the three denatured forms (30, 48, and 59 kDa proteins) of the permethrin CbE (CbE) from CPB as described by Lee and Clark (19). Each antiserum showed cross-reactivities to all the denatured forms of CbE as judged by immunoassay. An antibody capture immunoassay was developed using the immunoglobulin purified from the 30

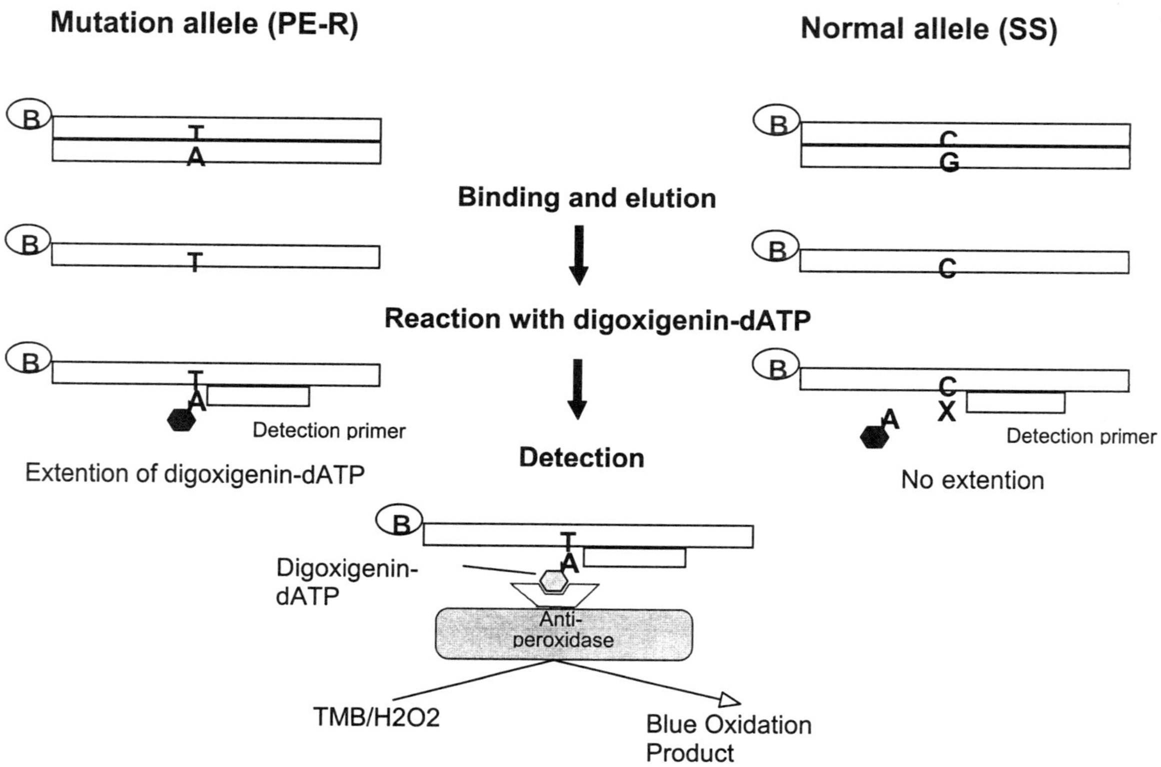

Figure 2. Schematic diagram of minisequencing reaction for the detection of the *LdVssc1* L1014F (C to T) mutation.

kDa antiserum in conjunction with denatured hemolymph as the antigen source. In this antibody capture immunoassay, the hemolymph from the permethrin-resistant strain produced 1.1-5.3 times higher signal levels than that from the susceptible (SS) strain depending on assay conditions, suggesting that the hemolymph from the resistant strain contains more CbE. These findings corroborated our original findings that permethrin CbE is overexpressed in the hemolymph of the permethrin-resistant strain of CPB and contributes to overall permethrin resistance (19).

The relative quantities of CbE in four different strains of CPB with different levels of permethrin resistance were assessed with this immunoassay and compared with a carboxylesterase activity assay using α-naphthyl caproate (αNC) as substrate (11).

Analysis of Molecular Techniques

bi-PASA Analysis

Both homozygous and heterozygous genotypes for the S291G and L1014F mutations are efficiently identified on the basis of size differences of amplified DNA fragments as shown in Fig. 3. For the detection of the *AChE* S291G mutation by bi-PASA, homozygous resistant (see RR lanes) or susceptible alleles (see SS lanes) are determined by the presence of representative large (206 bp) and small (126 bp) DNA fragments, respectively (Fig. 3A). Heterozygous alleles (see SR lanes) are determined by the presence of both 206-bp and 126-bp fragments (Fig. 3A). Likewise, the homozygous resistant (see RR lanes), homozygous susceptible (see SS lanes), and heterozygous alleles (see SR lanes) for the *LdVssc1* L1014F mutation are detected by the presence of large (170 bp), small (98 bp), or both 170 bp and 98 bp DNA fragments, respectively (Fig. 3B). Allele-nonspecific fragments (293 bp in the S291G bi-PASA and 203 bp in the L1014F bi-PASA) are generally absent or faint in the homozygous samples. Low amplification levels are probably due to their larger size because such amplifications may be thermodynamically less favorable than those of shorter allele-specific fragments. In heterozygous sample, however, simultaneous amplification of the two allele-specific fragments becomes thermodynamically less favorable than the amplification of a single fragment, which allows relatively more amplification of larger allele-nonspecific fragments as shown in Fig. 3 (see SR lanes).

The performance of bi-PASA is more dependent on the thermal cycler program, particularly annealing temperature, than other factors such as Mg^{++} concentration (data not shown). Template DNA quality also appears to be a critical factor for ensuring definitive resolution of amplified DNA fragments by

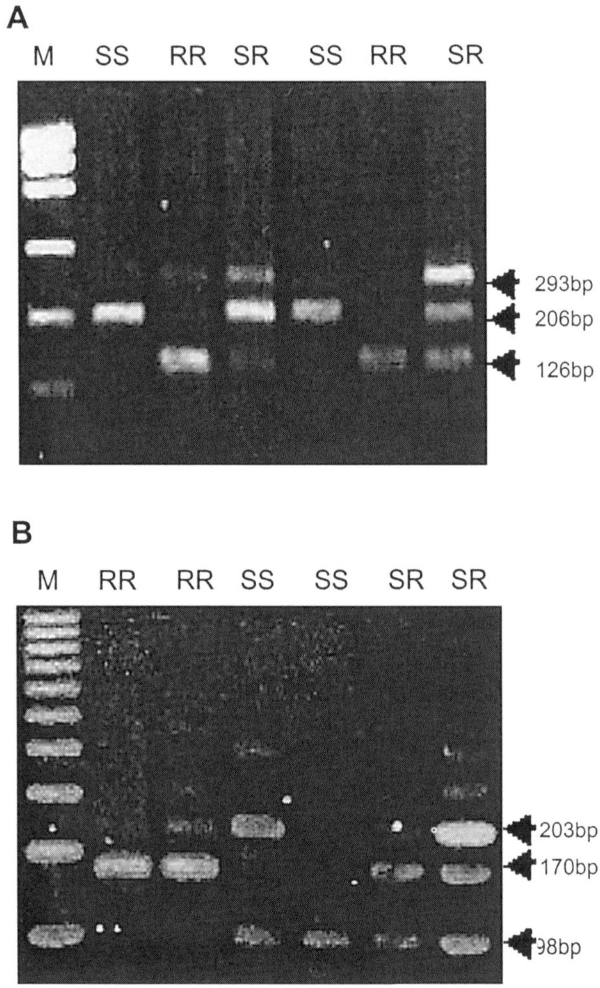

Figure 3. Typical bi-PASA banding patterns of the homozygous and heterozygous alleles of CPB for the *AChE* S291G (A) and *LdVssc1* L1014F (B) mutations. bi-PASA reactions using individual genomic DNA were separated on a 2% Metaphor agarose (FMC Bioproducts) gel and visualized by ethidium bromide. M, DNA marker; SS, susceptible homozygous; SR, heterozygous; RR, resistant homozygous.

bi-PASA since genomic DNA is used directly as template. Genomic DNA extracted from any stages of CPB larva or adult using DNAzol or DNAzol-BD (Molecular Research Center Inc.) is sufficient to generate unambiguous bi-PASA results. Results from bi-PASA are not always diagnostic when genomic DNA is extracted from eggs and used as template. bi-PASA profiles, nevertheless, are unaffected by template DNA concentrations over a range of 50 to 300 ng per PCR (data not shown).

Overall, the bi-PASA protocol is very rapid and simple. bi-PASA for 96 samples usually required approximately 2.5 hrs to complete, including sample handling. Agarose gel electrophoresis for 96 samples usually required 2.5 hrs, including gel preparation and sample loading time. Total cost for bi-PASA is calculated to be approximately $0.80 per sample, which is the lowest among the three methods evaluated. One intrinsic limitation in applying bi-PASA is that the success of bi-PASA appears to be mainly governed by the sequence context around mutation site. A too high G/C or A/T content surrounding the mutation site does not allow the design of effective allele-specific primers and accordingly a definitive distinction between two different alleles is not feasible. We have experienced difficulties in optimizing bi-PASA for the detection of *AChE* R30K mutation (2) in which the A/T content at the mutation region (20-base upstream and 20-base downstream from the mutation site) is very high.

Analysis of SSCP

SSCP also is very effective in identifying each genotype, including the heterozygous allele. As shown in Fig. 4, resistant homozygous alleles for both the *AChE* S291G and *LdVssc1* L1014F mutations (Fig. 4A and B, lane 3, respectively) are unambiguously distinguished from susceptible homozygous alleles by their unique banding patterns of single stranded DNA (Fig. 4A and B, lane 1). Heterozygous alleles appear as superimposed images of both homozygous susceptible and resistant banding patterns (Fig. 4A and B, lane 2).

For maximum resolution, it is critical to optimize electrophoresis conditions, including the running temperature and gel concentration. For the detection of *AChE* S291G mutation, electrophoresis at room temperature resulted in better band resolution than at low temperature. For the detection of *LdVssc1* L1014F mutation, however, a low running temperature (4° C) for electrophoresis is required to obtain sufficient band resolution between the two different alleles. A temperature-dependent resolution by SSCP, likewise, has been reported in the detection of a mutation in human HRAS1 gene (20).

SSCP is a very simple technique that is rugged and easily accomplished. Overall cost for SSCP is calculated to be approximately $1.20 per sample, which is slightly more expensive than bi-PASA but still is in a

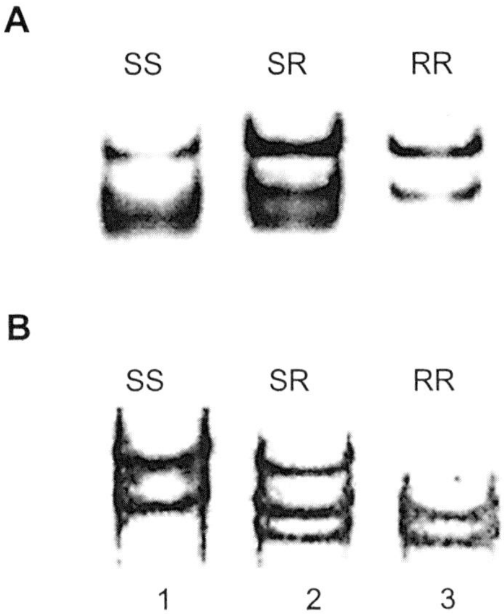

Figure 4. Representative SSCP patterns of the homozygous and heterozygous alleles of CPB for the *AChE* S291G and *LdVssc1* L1014F mutations. A. The *AChE* genomic DNA PCR fragments (247 bp) from the SS homozygous (lane 1), SS/AZ-R heterozygous (lane 2), and AZ-R homozygous (lane 3) individuals were separated on a 0.5 X MDE gel at room temperature. B. The *LdVssc1* genomic DNA PCR fragments (349 bp) from the SS homozygous (lane 1), SS/PE-R heterozygous (lane 2), and PE-R homozygous (lane 3) individuals were separated on a 0.5 X MDE gel at 4°C. DNA bands were detected by silver staining.

competitive range. SSCP, however, requires a longer time commitment. It usually required 2.5 hrs to run the initial PCR, 7 to 17 hrs for gel preparation and electrophoresis and 1 hr for silver staining. Additionally, the indirect nature of the mutation detection process (differential migration of single stranded DNA fragments) is considered to be another drawback. Obviously, silent mutations and polymorphisms not associated with resistance could be a source of false positive and negative results.

Analysis of Minisequencing

Recently, we have developed a minisequencing reaction scheme that is coupled with a 96-well microtiter plate format, which allows direct detection of mutations associated with resistance using a high throughput assay. The results from the minisequencing reaction for the detection of *AChE* S291G mutation has been previously described (16) (Fig. 5A). Briefly, high levels of absorbance (OD_{dATP} = 0.319 ± 0.05) are obtained from individual beetles with the homozygous susceptible allele (SS strain) due to the presence of unmutated wildtype base T at the mutation site, to which digoxigenin-dATP is incorporated. In contrast, approximately 10-fold lower absorbance levels (OD_{dATP} = 0.031 ± 0.018) are detected from individual beetles with the homozygous resistant allele (AZ-R strain, mutated base C at the mutation site). Based on this highly significant difference in OD values, unambiguous identification of beetles with homozygous resistant allele is possible.

In the minisequencing reaction for the detection of *LdVssc1* L1014F mutation, a biotinylated sense strand of the 349 bp genomic DNA fragment is immobilized onto a streptavidin-coated 96-well plate and the detection primer is allowed to anneal one base immediately downstream of the mutation site (C or T), leaving the 3' end of the annealed detection primer available for the incorporation of digoxigenin-dATP or -dUTP. As illustrated in Fig. 5B (see PE-R), higher absorbance readings (OD_{dATP} = 0.312 ± 0.022) are observed due to the incorporation of digoxigenin-dATP to the base T at the mutation site (L1014F, C to T base substitution resulting in the homozygous resistant allele). For the homozygous susceptible allele (see SS, Fig. 5B), approximately 10-fold lower absorbance levels (OD_{dATP} = 0.038 ± 0.018) are obtained due to the presence of the unmutated base C at the mutation site, which does not allow the incorporation of digoxigenin-dATP to the DNA template. These values are not significantly different from the nonspecific signal levels as measured by OD_{dUTP} values (0.043 ± 0.023 and 0.031 ± 0.018 for the homozygous susceptible and resistant alleles, respectively; *t*-test, $P > 0.25$). The significant difference in the specific OD_{dATP} values between resistant and susceptible alleles (*t*-test, $P < 0.001$) allows an unambiguous detection of the resistant mutation in individual

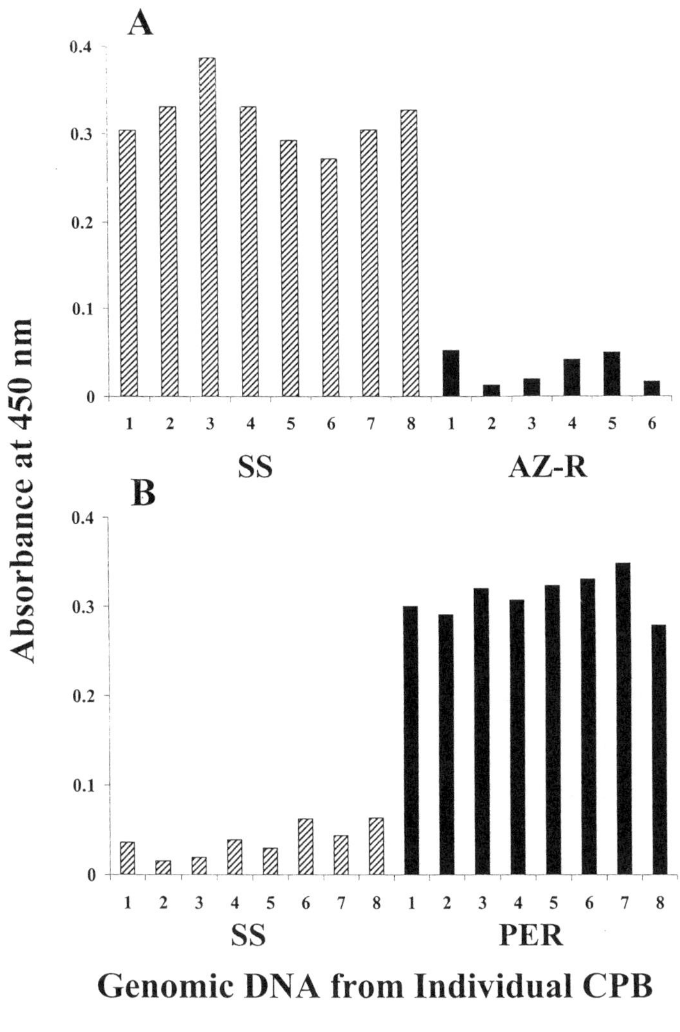

Figure 5. Typical results shown in OD values from the minisequencing reactions for the detection of the *AChE* S291G (A, adapted from ref [13]) and *LdVssc1* L1014F (B) mutations. Homozygous susceptible (SS) and resistant (AZ-R and PE-R) DNA samples were analyzed by minisequencing.

PE-R CPB. Preliminary results from the minisequencing of heterozygous alleles (SR) suggested that specific OD_{dATP} values from the standard heterozygous DNA samples are intermediate between those from the homozygous resistant and susceptible alleles but more analyses of a large number of samples will be required to obtain a statistically reliable range of values for their unambiguous detection.

Minisequencing is a relatively quick protocol (approximately 4 hrs) and is a direct mutation detection method in which the presence or absence of specific base at a mutation site can be measured as distinct absorbance values. Compared to bi-PASA and SSCP, the minisequencing reaction using a 96-well format can analyze a large number of samples (up to 96 samples) simultaneously and, if equipped with a robotic liquid handling system, may be adapted for a high throughput analysis. Nevertheless, it is relatively more labor-intensive protocol, requiring longer time commitment and more technical expertise of the experimenter. The cost for minisequencing of 96 samples is calculated as approximately $500 (ca. $5 per sample), which is the most expensive among the three tested but still cheaper than automatic sequencing. Lastly, the minisequencing protocol is limited by the availability of digoxigenin-labeled nucleotides necessary for sequencing. At this time, since only digoxigenin-dUTP and digoxigenin-dATP are available, the detection of C to G or G to C point mutations is not feasible by minisequencing.

Analysis of Antibody Capture Immunoassay

In order to determine the optimum conditions for the antibody capture immunoassay, various combinations of hemolymph dilution (denatured hemolymph from the SS and PE-R strains, 10^1 to 10^5-fold dilution, 5.5 ng-55 µg total protein/well) and 1° antibody (AB) dilution (IgG purified from 30 kDa antiserum, 50 to 250-fold dilution, 100 ng-2.5 µg 1°AB/well) were tested. In considering several factors including signal specificity, signal intensity, and convenience of hemolymph dilution, the combination of 100-fold dilution of hemolymph (5.5 µg denatured total protein/well) and 250-fold dilution of 1°AB (500 ng/well) was determined to be the optimum condition for the detection of CbE proteins in CPB hemolymph (19).

To test the sensitivity of the antibody capture immunoassay for the detection of the permethrin CbE from individual CPB, hemolymph samples from 41-47 individual larva of 4 different CPB strains were comparatively analyzed using the CbE activity assay and the antibody capture immunoassay. The CbE activity of the hemolymph of individual larvae were determined by using αNC as a substrate and the results presented in the order of high αNC hydrolysis activity to low (Fig. 6, panel A). The levels of CbE in the hemolymph of individual

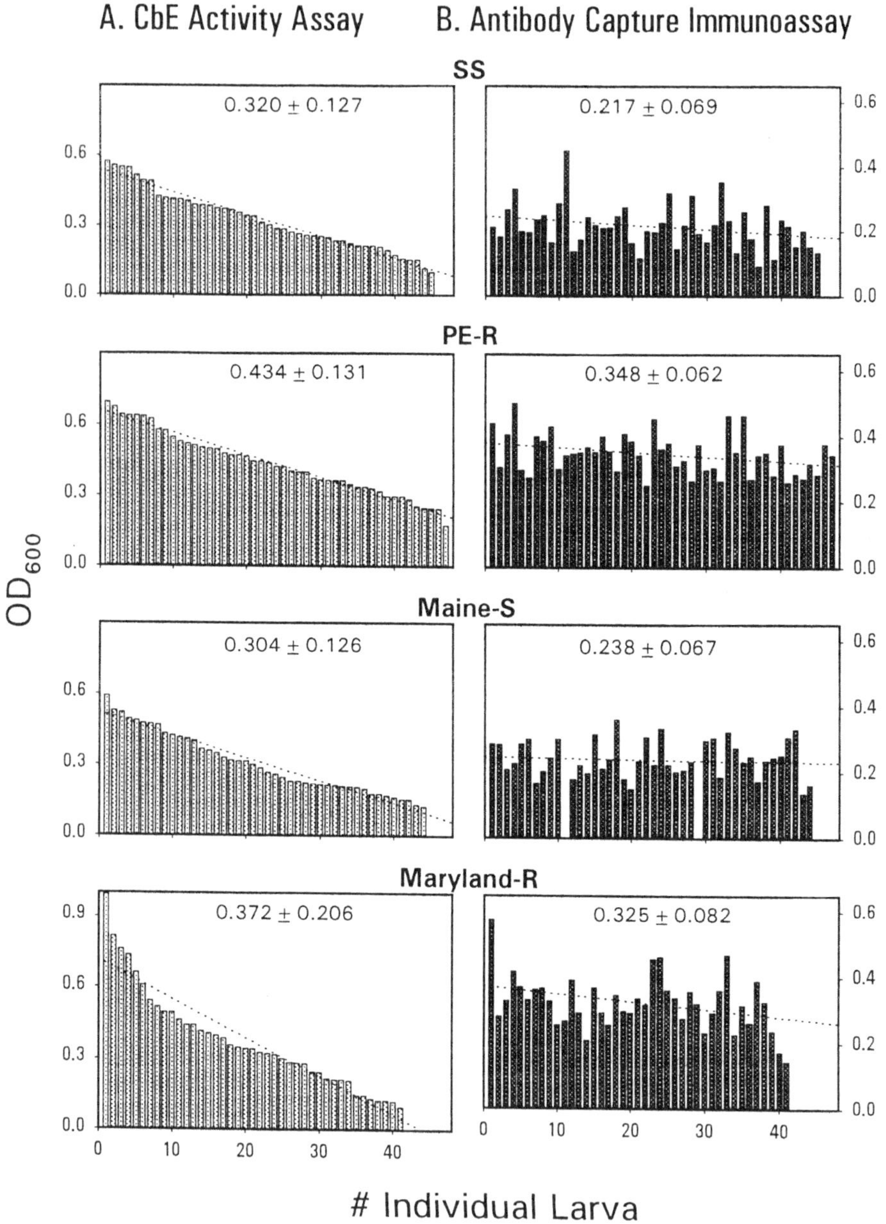

Figure 6. Carboxylesterase (CbE) activity assay (A) and antibody capture immunoassay (B) of individual hemolymph samples from four different strains of CPB. Individual larvae were sorted by the order of the CbE activity from high to low. The value in each panel is the mean ± standard deviation (S.D.) of 41-47 separate determinations (Reproduced with permission from reference 19. Copyright 1999, Academic Press).

larvae were determined by the antibody capture immunoassay and the results presented according to the individual larval number sorted by the CbE activity (Fig. 6, panel B). In the CbE activity assay, the resistant PE-R strain showed the highest mean αNC CbE activity (0.434 $\pm$ 0.131 OD_{600}) and followed by the Mar-R (0.372 $\pm$ 0.206 OD_{600}), SS (0.320 $\pm$ 0.127 OD_{600}), and Mai-S (0.304 $\pm$ 0.126 OD_{600}) strains (Fig. 6, panel A). The mean αNC CbE activity was statistically elevated in the PE-R strain when compared with the SS and Mai-S strain (*t*-test, n=90-91, $P<0.05$). The PE-R, Mar-R, and Mai-S strains had 1.4, 1.2, and 1.0 times higher levels of the αNC CbE activity when compared with the SS strain based on mean values. Since the Mar-R strain has a similar level of permethrin resistance compared with the PE-R strain and the Mai-S strain has similar level of permethrin-susceptibility compared with the SS strain (unpublished observations, R. Alford, G. Dively, and S. H. Lee), the levels of α NC CbE activity correlate to a considerable degree with the levels of permethrin resistance. The Mar-R strain had a relatively high degree of variation in αNC CbE activity among individuals when compared with the other strains as judged by standard deviation of αNC hydrolysis (0.206 *vs.* 0.126-0.131 S.D., respectively). Due to this high variation among individuals, the elevated level of αNC CbE activity of the Mar-R strain could not be determined to be significantly different either from that of the PE-R and SS strains, respectively (*t*-test, n=85-88, P=0.094-0.156). However, the mean value of αNC CbE activity and bioassay data confirm that the strain is permethrin-resistant.

In the antibody capture immunoassay, the PE-R strain showed the highest level of CbE (0.348 $\pm$ 0.062 OD_{450}) followed by the Mar-R (0.325 $\pm$ 0.082 OD_{450}), Mai-S (0.238 $\pm$ 0.067 OD_{450}), and SS (0.217 $\pm$ 0.069 OD_{450}) strains (Fig. 4, panel B). Both the PE-R and Mar-R strains were determined to have significantly higher levels of the CbE than the SS or Mai-S strains (*t*-test, n=83-91, $P<0.05$). The PE-R, Mar-R, and Mai-S strains had 1.6, 1.5, and 1.1 times higher levels of the CbE when compared with the level of the permethrin CbE in the SS strain based on mean values. In terms of inter-strain comparisons, therefore, the average levels of the CbE, determined by the antibody capture immunoassay, were better correlated with the levels of permethrin resistance than the αNC CbE activity determinations. This result indicates that the antibody capture immunoassay is more precise and specific in evaluating the mean levels of the permethrin CbE in CPB populations than the CbE activity assay.

In a pair-wise comparison between the level of the CbE and the αNC CbE activity in each individual within a strain, both factors were positively correlated but with very low correlation coefficients (r^2=0.0648, 0.0966, 0.0084, and

0.1721 in the SS, PE-R, Mai-S, and Mar-R strains, respectively: also see the regression lines in Fig. 6). This low correlation primarily appears to be due to the relatively low specificity intrinsically associated with the CbE activity determination using a non-specific substrate such as αNC (9, 10). This finding also implies that the level of CbE activity in each individual as a marker for the resistance mediated by CbE may result in considerable amount of error in predicting permethrin resistance in CPB.

Conclusions

The advantages and disadvantages for each of the DNA-based genotyping methods for the detection of resistance mutation are summarized in Table 2. Overall, a combination of techniques appears necessary for unambiguous genotyping in that no single technique is suitable for the detection of all mutations. Based on these considerations, we have established an efficient genotyping strategy for the detection of resistance mutations in the CPB. For efficient and rapid sampling of CPB populations, it is recommended to collect egg masses rather than other stages that require more extensive sample handling and care. However, the extraction of genomic DNA from individual eggs is not efficient due to the large amount of lipid and yolk protein and relatively small amount of DNA in eggs, resulting in poor quality DNA that is not suitable for bi-PASA and other DNA amplification techniques. An alternative strategy to overcome this problem is to extract genomic DNA from newly hatched neonates and we have determined that the quality of neonate genomic DNA is sufficient and satisfactory for bi-PASA and the other DNA-diagnostic techniques (data not shown). Genomic DNA is preferred as templates for the detection of resistance mutations since it allows the detection of heterozygous genotype and is significantly more stable and easier to obtain and use than cDNA. Once template genomic DNA is available, we suggest bi-PASA as the primary diagnostic method due to its ruggedness, rapidity, simplicity, cost effectiveness, and accuracy. SSCP and minisequencing are amenable as secondary methods to verify any ambiguous primary results obtained from the initial bi-PASA. Finally, it is recommended to conduct random direct sequencing of one in 50 or 100 samples for QA/QC purposes. Availability of these DNA-based genotyping techniques using neonate genomic DNA will allow the rapid and precise monitoring of the resistant and susceptible allele frequencies, including heterozygous type, in the field populations of CPB.

Finally, the antibody capture immunoassay is more reliable in evaluating the level of permethrin CbE in the hemolymph of CPB compared to the CbE activity assay. In spite of its relatively more complicated procedures, the antibody capture immunoassay is expected to be an effective and sensitive means to detect

Table II. Advantages and disadvantages of the three DNA-based genotyping methods for the detection of the *AChE* S291G (A) and *LdVssc1* L1014F (B) mutations associated with resistance

Method	Advantage	Disadvantage
bi-PASA	1. Simple, easy and rapid 2. Efficient genotyping 3. Easy to process a large number of samples 4. Direct detection of mutation	1. Dependent on sequence context around mutation site 2. Sensitive to template DNA quality
SSCP	1. Simple and easy 2. Efficient genotyping 3. Detection of allelic variations	1. Longer time commitment 2. Difficult to process a large number of samples 3. Indirect detection of mutation
Minisequencing	1. Adaptable to high throughput analysis 2. Relatively rapid 3. Direct detection of mutation	1. Relatively less clear genotyping 2. Labor intensive 3. More expensive

the different levels of permethrin CbE in resistant and susceptible CPB populations and can be used to corroborate results obtained enzymatically. Additionally, the purified IgG is now available for large scale affinity purification of permethrin CbE for microsequencing analysis to obtain degenerate RT-PCR primers for the cloning and sequencing of the *permethrin CbE* gene.

Acknowledgment

This work was supported by research grants, USDA (NAPIAP)-TPSU-UM-3361-529, MAES 617), USDA Regional Research Grant NE-180 and USDA(NRICGP) 95-37302-1887.

References

1. Argentine, J.A.; Clark, J.M.; Ferro, D.N. *J. Econ Entomol.* **1989,** 82, 698-705.
2. Dunn, J.B. M.S. Thesis, University of Massachusetts-Amherst. **1999,** 1- 93.
3. Argentine, J.A.; Zhu, K.Y.; Lee, S.H.; Clark, J.M. *Pestic. Biochem. Physiol.* **1994,** 48, 63-78.
4. Zhu, K.Y.; Clark, J.M. *Insect Biochem. Mol. Biol.* **1995,** 25, 1129-1138.
5. Zhu, K.Y.; Lee, S.H.; Clark, J.M. *Pestic. Biochem. Physiol.* **1996,** 55, 100-108.
6. Zhu, K.Y.; Clark, J.M. *Pestic. Biochem. Physiol.* **1995,** 51, 57-67.
7. Argentine, J.A.; Lee, S.H.; Sos, M.A.; Barry, S.R.; Clark, J.M. *Pestic. Biochem. Physiol.* **1995,** 53, 97-115.
8. Lee, S.H.; Dunn, J.B.; Clark, J.M.; Soderlund, D.M. *Pestic. Biochem. Physiol.* **1999,** 63, 63-75.
9. Lee, S.H.; Clark, J.M. *Pestic. Biochem. Physiol.* **1996,** 56, 208-219.
10. Lee, S.H.; Clark, J.M. *Pestic. Biochem. Physiol.* **1998,** 60, 31-47.
11. Lee, S.H.; Clark, J.M. *Pestic. Biochem. Physiol.* **1998,** 62, 51-63.
12. Williamson, M.S.; Martinez-Torres, D.; Hick, C.A.; Castells, N.; Devonshire, A.L. In *Molecular Genetics and Evolution of Pesticide Resistance;* Brown, T. M., Ed.; American Chemical Society: Washinton, DC, 1996; pp 52-61.
13. Liu, Q.; Thorland, E.C.; Heit, J.A.; Sommer, S.S. *Genome Res.* **1997,** 7, 389-398.
14. Coustau, C.; ffrench-Constant, R. *Pestic. Sci.* **1995,** 43, 267-271.
15. Honyo,T.; Buzard, G.S.; Calvert, R.J.; Weghorst, C.M. *Nucleic Acids Res.* **1993,** 21, 3637- 3642.
16. Zhang, A.; Dunn, J.B.; Clark, J.M. *Pestic. Biochem. Physiol.* **1999,** 65, 25-35.
17. Switzer, R.C.; Merril, C.R.; Shifrin, S. A. *Anal. Biochem.* **1979,** 98, 231-237.

18. Syvanen, A.C.; Ikonen, E.; Manninen, T.; Bengtstrom, M.; Soderlund, H.; Aula, P.; Peltonen, L. *Genomics* **1992,** 12, 590-595.
19. Lee, S.H.; Clark, J.M. *Pestic. Biochem. Physiol.* **1999,** 64, 66-75.
20. Orita, M.; Iwahana, H.; Kanazawa, H.; Hayashi, K.; Sekiya, T. *Proc. Natl. Acad. Sci. U S A* **1989,** 86, 2766-2770.

Herbicide Resistance

Chapter 7

Auxinic Herbicide Resistance in Wild Mustard (*Sinapis arvensis* L.)

J. Christopher Hall, Nataraj N. Vettakkorumakankav, and Hong-gang Zheng

Department of Environmental Biology, University of Guelph, Guelph, Ontario N1G 2W1, Canada

Plant species resistant to auxinic herbicides have been identified and the basis for this resistance remains unclear. Our objectives are to understand the molecular mechanism(s) of auxinic herbicide resistance in wild mustard (*Sinapis arvensis* L.). In order to achieve this objective, we characterized both a susceptible (S) and resistant (R) biotype of wild mustard at the biochemical, physiological and molecular levels (Table 1). When the S and R biotypes were compared, there were no differences in the uptake, transport and metabolism of auxinic herbicides. However, significant differences in auxinic herbicide-induced ethylene biosynthesis were observed between the two biotypes. We have recently studied the role of auxinic herbicide-induced ethylene on root morphology. Biophysical studies indicate an important role for calcium in mediating the resistance phenomenon. Furthermore, molecular analysis of auxin-binding proteins (ABPs) from both biotypes revealed important differences in the structure of this putative auxin receptor, which may be crucial for the development of resistance.

 © 2002 American Chemical Society

Table I. Characteristics of the auxinic herbicide-resistant (R) and -susceptible (S) biotypes of *S. arvensis*

S biotype	*R biotype*
Susceptible to auxinic herbicides	Highly resistant to picloram and dicamba, moderately resistant to 2,4-D and MCPA and little or no resistance to MCPP and 2,4-DP.
Normal Morphology	Shorter plants, more branched with darker green leaves having more chlorophyll and cytokinins
Auxinic herbicides induce denovo biosynthesis of ACC synthase and ethylene	Auxinic herbicides have little or no effect on the biosynthesis of ACC synthase and ethylene
Two IAA-binding activities (high and low affinity) in ABP preparations	One IAA-binding activity in ABP preparations
Picloram-induced proton efflux in isolated protoplasts	No picloram-induced proton efflux in isolated protoplasts
Only one phase swelling was observed when herbicide-induced protoplastic volume changes were measured by light scattering spectroscopy	Two phases, a shrinking and swelling phase, were observed
Calcium ionophore A23187 plus calcium reduce the sensitivity of protoplasts to auxinic herbicides	Calcium channel blocker (verapamil) increases the sensitivity of protoplasts

Plant growth and development, from germination to senescence, is controlled by the interaction between phytohormones including auxins, abscisic acid (ABA), cytokinins, ethylene and gibberellins. Many compounds with growth regulating activities have been synthesized for modulating plant growth and development and include the auxinic herbicides. Auxinic herbicides are so named because they cause many of the same responses as IAA in plants. They include an increase in ethylene evolution following herbicide application, a dose-dependent growth response, ability of auxinic herbicides to replace IAA in tissue culture, as well as their ability to cause cell elongation and other physiological changes (for a detailed review see 1). Auxinic herbicides were the first group of organic herbicides that revolutionized modern agriculture and have remained in use for over fifty years for the control of broadleaf weeds in cereal crops (2). These herbicides include members from different groups such as phenoxyalkanoic acids (e.g. 2,4-D, MCPA, MCPP), benzoic acids (e.g. dicamba, chloramben), and pyridines (e.g. picloram, triclopyr, clopyralid) and may also include the quinolinic acids. Changes in response to auxinic herbicides include cell elongation, epinasty, hypertrophy, root initiation, ethylene and cyanide biosynthesis that ultimately result in the death of susceptible species (2,3). Additionally, a rapid Ca^{2+} influx, activation of plasma membrane ATPase, and increased nucleic acid and protein biosynthesis are also observed. All these changes are believed to be elicited via the interaction of auxinic herbicides with putative receptor(s) such as auxin-binding proteins (ABPs), triggering a cascade of events that ultimately lead to the observed morphological symptoms. The precise nature of the putative auxinic receptor as well as the nature of these molecular signals remain to be ascertained.

The prolonged use of auxinic herbicides has led to the development of resistance in a variety of species (4-9) including wild mustard (*Sinapis arvensis* L.) which is studied in our laboratory. This auxinic herbicide resistant biotype was identified in a field in Western Canada that had been repeatedly treated with a combination of 2,4-D/dicamba/MCPA for over 10 years (6,10,11). Classical genetic analysis indicates that this auxinic herbicide resistance is controlled by a single, completely dominant nuclear allele (11). The objective of the research conducted in our laboratory is to understand the mode of action of auxinic herbicides as well as to exploit the potential of auxinic herbicide resistance in modern agriculture. Our understanding of mechanism of auxinic herbicide action has been reviewed recently (1) and, in this paper, we will limit our discussion to auxinic herbicide resistance in wild mustard.

Morphological and Physiological Characteristics of Auxinic Herbicide-Resistant Wild Mustard

Compared to the susceptible (S) biotype of wild mustard, the resistant (R) biotype is shorter and more branched (12). However, there are no differences in fresh or dry weights between the biotypes. Plants of the R biotype have more leaves that are significantly smaller and darker green, and have a less developed root system compared to the S biotype. The R biotype is more cold tolerant (-7°C) at the four-leaf stage of development with higher germination rates at 30°C. Furthermore, the R biotype has elevated levels of cytokinins, suggesting that the hormonal balance may be affected in this biotype (12).

In order to determine the basis for the observed resistance to auxinic herbicides in wild mustard, the absorption, translocation and metabolism of auxinic herbicides were compared between the S and R biotypes (5). These studies reveal that uptake and translocation of ^{14}C-radiolabelled 2,4-D, dicamba and picloram are similar in both biotypes. Similarly, there are no differences in rates of metabolism of these herbicides in both biotypes (5). Treatment of both biotypes with standard levels of picloram results in epinasty of the S biotype within 24 hours, whereas a similar effect is not observed in the R biotype (13).

Ethylene Biosynthesis

There are differences in biosynthesis of ethylene and its precursors between S and R biotypes in response to auxinic herbicide treatment (6,13). The S biotype produces two- and six-fold more ethylene than the R biotype within 4 and 44 h of treatment with auxinic herbicides (13). The significance of this difference in levels of auxinic herbicide-induced ethylene biosynthesis between the two biotypes is further supported by the fact that fumigation of the R biotype with ethylene results in auxinic herbicide-like epinasty. This observation led to the suggestion that differences in auxinic herbicide-induced ethylene may be involved in mediating auxinic herbicide resistance (13). It has also been demonstrated that in the S biotype increased biosynthesis of ethylene in response to picloram treatment is regulated by *de novo* synthesis of 1-aminocyclopropane-1-carboxylic acid synthase (ACC synthase; ACS) while in the R biotype the levels of ACS activity are not altered in response to picloram (13). The fact that auxinic herbicide-induced ethylene plays a crucial role in eliciting the observed morphological and physiological changes in plants is well documented in the literature (14-18).

In order to delineate the role of auxinic herbicide-induced ethylene biosynthesis in the development of morphological symptoms following auxinic herbicide treatment, we identified and characterized the *ACS* gene family in wild mustard (S biotype) and studied their auxin-induced gene expression (19). Furthermore, the effects of auxinic herbicides on the growth and development of S wild mustard as well as wild-type, auxin-resistant and ethylene-insensitive lines of *Arabidopsis thaliana* were investigated. Following treatment with the auxinic herbicide 2,4-D typical morphological symptoms such as expansion of the radius of the root/hypocotyl and inhibition of root and hypocotyl elongation are observed in seedlings of S wild mustard and wild-type *Arabidopsis*. It was concluded that all of these morphological changes, except for inhibition of root elongation, are caused by auxinic herbicide-induced ethylene biosynthesis (19). This increase in ethylene biosynthesis is caused by transcriptional activation of auxin-inducible genes encoding ACS. This is the first report suggesting that two types of signals/mechanisms exist;

the first results from auxinic herbicide-induced ethylene (root/hypocotyl radial expansion and inhibition of hypocotyl elongation) while the second mechanism (inhibition of root elongation) does not appear to be mediated by ethylene. Additional studies on characterizing the *ACS* gene family in the R biotype as well as regulation of their expression are currently underway in our laboratory.

A Role for Calcium in Mediating Auxinic Herbicide Resistance

We have observed that the endogenous levels of calcium in the R biotype are 2 to 3 times higher than in the S biotype (unpublished results) and that cytokinin levels in the R biotype are also elevated (12). The elevated levels of both calcium and cytokinins and their interaction may play a role in mediating auxinic herbicide resistance. Indeed, such an interaction between calcium and cytokinins in delaying senescence has been previously reported (20, 21). Furthermore, in these studies (20, 21) it was observed that cytokinin was ineffective in delaying senescence in the absence of calcium and that this effect of cytokinin was restored in the presence of calcium. It has also been reported that cytokinin-induced bud formation is dependent on calcium (22). We have recently investigated the role of calcium in mediating auxinic herbicide resistance using light scattering spectroscopy.

Light scattering spectroscopy is a non-invasive probe that allows the detection of morphological changes in real-time (23, 24). Two types of light scattering signals were measured in these experiments (i) a nucleotide-independent flash-induced signal and (ii) an ATP-dependent flash-induced signal. Results indicate that the ATP-dependent signal, manifested as a conformational change, differs between the two biotypes (25). For example, the signal from R protoplasts remains unaffected by 50 μM picloram whereas the signal from the S protoplasts is reduced by 20 μM picloram and completely abolished by 50 μM picloram. The inclusion of the calcium channel blocker verapamil with picloram reduced the signal amplitude from R protoplasts. Conversely, in S protoplasts the inclusion of the calcium ionophore A23187 plus calcium reduced the inhibitory effects of picloram. Investigations into the efflux of protons from both biotypes using acridine orange as a pH indicator suggests that auxinic herbicide-induced proton efflux occurs only in the S biotype. These studies (25) suggest that auxinic herbicides change the calcium and hydrogen ion dynamics in the S biotype but not in the R biotype.

A role for calcium in mediating both auxinic herbicide action as well as resistance was further supported by Deshpande and Hall (26). ATP-dependent auxin and auxinic herbicide-induced volume changes in R and S protoplasts were investigated using right-angle light scattering. The scattering profile of IAA-induced volume changes consisted of an initial shrinking phase followed by a swelling phase, which is observed in both biotypes. Kinetic analysis of the swelling phase suggests two first-order rate processes for the S biotype, whereas only a single first-order rate process is observed in the R biotype. In the presence of the auxinic herbicide picloram, only a swelling phase is observed in the S biotype whereas, both shrinking and swelling of protoplasts are observed in the R biotype. Upon incubation of the R biotype with other auxinic herbicides such as MCPA or dicamba, both of which are effective to varying degrees on this biotype, the scattering profile is similar to that obtained in the S biotype with picloram treatment. Simultaneous incubation of R protoplasts with the calcium channel blocker verapamil reduces the initial auxinic herbicide-induced shrinking phase, once again supporting a role for calcium in mediating auxinic herbicide resistance (26).

The role of calcium in mediating resistance to auxinic herbicides at the level of intact seedlings has also been investigated by modulating calcium dynamics of S and R seedlings of wild mustard (27). The inhibitory effects of the auxinic herbicides on the S biotype are significantly reduced upon pre-treatment of seedlings with calcium in the presence of the calcium ionophore A23187. Conversely, the addition of a calcium channel blocker verapamil to the R biotype increases its sensitivity to the auxinic herbicides. However, the potassium channel ionophore valinomycin does not have the same effects on seedlings of wild mustard treated with auxinic herbicides, suggesting that the observed effects are specific for calcium. These *in vivo* results are consistent with those from our earlier *in vitro* experiments with isolated protoplasts, and demonstrate, for the first time, that calcium plays a crucial role in mediating auxinic herbicide resistance in wild mustard at the level of intact seedlings.

Characterization of ABP from Wild Mustard

There is growing evidence that ABP is a functional auxin receptor (28). Based on initial studies, Webb and Hall (29) hypothesized that there may be inherent differences in ABP between the S and R biotypes of wild mustard. ABP preparations from both the biotypes have similar substrate (^{3}H-IAA) binding and time course profiles. Scatchard analysis of ^{3}H-IAA binding to ABP preparations from the S biotype indicates there are two populations of binding sites; a low and a high affinity-binding site whereas, the R biotype has only one binding site, which is comparable to the low affinity-binding site in the S biotype. These results are supported by light scattering data indicating two first-order rate processes in the swelling phase of the S biotype (while only one first-order rate process is observed in the R biotype) in response to auxinic herbicide treatment (26). These results prompted us to suggest differences in auxin receptor(s) as a mechanism in mediating resistance to auxinic herbicides in wild mustard.

To further investigate this difference in auxin/auxinic herbicide binding, the effects of IAA on auxinic herbicide-induced changes in proton dynamics were monitored by observing the nucleotide-independent flash induced signals in protoplasts from both biotypes (30). Preincubation of S protoplasts with IAA does not affect the auxinic herbicide-induced proton efflux, whereas preincubation with IAA reduces the auxinic herbicide-induced proton efflux from R protoplasts, thereby moderating the deleterious effects of the herbicides. It is possible that an allosteric interaction model could describe the kinetics of the observed interaction between IAA and auxinic herbicide in the R biotype. In other words, the interaction of IAA with the receptor protein molecule may cause a conformational change which may affect the ability of the herbicide to bind to the receptor, thus reducing the herbicidal effect in the R biotype. This phenomena does not occur in the S biotype (30).

Estelle and Sommerville (31) identified several 2,4-D resistant *Arabidopsis* lines. This resistance is controlled by a single recessive gene (*AXR1*) that encodes a protein related to ubiquitin-activating enzyme E1 (32). Together with other components of the ubiquitin system, *AXR1* regulates the stability of a wide range of proteins involved in an auxin signaling pathway (33). Similar to our observations with *S. arvensis*, the resistance to auxinic herbicides in these resistant lines of *Arabidopsis* was not due to differences in uptake or metabolism of the herbicide. These resistant lines of *Arabidopsis* were shorter, bushy with small, thin roots (31). Subsequent research on 20 *axr* 1 mutants of *Arabidopsis* identified at least five *axr* 1 alleles, but all these mutants exhibited similar phenotypes (34). The morphological characteristics exhibited by the *axr* 1 mutants of *Arabidopsis* are remarkably similar to those observed in the R biotype of *S. arvensis*. Furthermore, it has been demonstrated that dicamba resistance in wild mustard is determined by a single, completely dominant nuclear allele (11).

Recently, we have isolated and characterized a gene family encoding ABP from the R and S biotypes of wild mustard (Zheng and Hall, unpublished). Differences in the nucleotide-derived amino acid sequence were observed that might account for auxinic herbicide resistance. Functional analysis of cDNAs encoding ABP is underway to determine the role of ABP in mediating auxinic herbicide resistance.

Conclusions

Signal transduction pathways mediating auxin response(s) have been extensively investigated. We have studied auxinic herbicide resistance and mode of action at the biochemical, physiological and molecular levels using an auxinic herbicide-resistant biotype of wild mustard as our model. Our results have provided a better understanding of the mode of action of auxinic herbicides as well as the basis of resistance to these herbicides, and we hypothesize that an interaction between various phytohormones and calcium may play significant role in these two phenomena. Discovery of molecular mechanisms of auxin action will greatly depend on the cloning and functional analysis of auxin receptor genes and auxin-induced genes involved in the auxin signaling pathways. Molecular genetic approaches will be an efficient way to discover these molecular targets of auxin/auxinic herbicides as well as their functions. Molecular marker-based cloning is another strategy that can be used to clone genes conferring auxinic herbicide resistance in wild mustard. This research will enhance our understanding of the molecular basis of the mode of auxin action and may lead to its application in agriculture via creating auxinic herbicide resistance crops.

References

1. Sterling, T. M.; Hall, J. C. In *Herbicide Activity: Toxicology, Biochemistry and Molecular Biology*; Roe, R. M. Ed.; IOS Press 1997; pp 111-141.
2. Devine, M. D.; Duke, S. O.; Fedtke, C. In *Physiology of Herbicide Action*; Prentice Hall Inc., Englewood Cliffs, NJ, 1993; pp. 295-309.
3. Coupland, D. In *Herbicide Resistance in Plants*; Powles, S. B.; Holtum, J. A. M. Eds.; Lewis Publishers, Boca Raton, FL, 1994; pp 171-214.
4. Barnwell, P.; Cobb, H. *Weed Res.* **1989**, *29*, 136-140.
5. LeBaron, H. M.; Gressel, J. *Herbicide Resistance in Plants*; Wiley Publishers: New York, NY, 1992; 401pp.
6. Peniuk, M. G.; Romano, M. L.; Hall, J. C. *Weed Res.* **1993**, *33*, 431-440.
7. Migo, T. R.; Mercado, B. L.; De Datta, S. K. Philippine, J. *Weed Sci.* **1986**, *13*, 28-38.
8. Bourdot, G. W.; Harrington, K. C.; Popay, A. I. *Proc. Brit. Crop Prot. Conf.-Weeds.* **1989**, *1*, 309-316.
9. Bourdot, G. W.; Hurrell, G.; Saville, D. J. *Weed Res.* **1990**, *38*, 449-457.
10. Heap, I.; Morrison, I. N. *Weed Sci. Soc. Amer. Abstr.* **1992**, *32*, 164.
11. Jasieniuk, M.; Morrison, I. N.; Brule-Babel, A. *Weed Sci.* **1995**, *43*, 192-195.
12. Hall, J. C.; Romano, M. L. *Pestic. Biochem. Physiol.* **1995**, *52*, 149-155.
13. Hall, J. C.; Alam, S. M. M.; Murr, D. P. *Pestic. Biochem. Physiol.* **1993**, *47*, 36-43.
14. Hall, J. C.; Soni, M. *Z. Naturforsch.* **1991**, *46*, 957-962.
15. Abeles, F. B. *Weed Sci.* **1968**, *16*, 498-500.
16. Holm, R. E.; Abeles, F. B. *Planta* **1968**, *78*, 293-297.
17. Morgan, P. W. In *Herbicides*; Audus, L. J. Ed.; Academic Press: New York, NY, 1976, Vol. 1; pp. 256-280.
18. Hall, J. C.; Bassi, P. K.; Spencer, M. S.; Vanden Born, W. H. *Plant Physiol.* **1985**, *39*, 18-23.
19. Wei, Y.D.; Zheng, H.G.; Hall, J. C. *Pest Manag. Sci.* **2000,** 56, 377-387.
20. Poovaiah, B. W.; Leopold, A. C. *Plant Physiol.* **1973**, *52*, 236-239.
21. Leopold, A. C.; Poovaiah, B. W.; Dela Fuente, R. K.; Williams, R. J. *Plant Growth Substances 1973*; Hirokawa Press; Tokyo, Japan, 1973; pp. 780-788.
22. Poovaiah, B. W.; Reddy, A. S. N. *CRC Crit. Rev. Plant Sci.* **1987**, *6*, 47-103.
23. Latimer, P. *Ann. Rev. Biophys. Bioeng.* **1982**, *11*, 129-150.
24. Latimer, P.; Moore, D. M.; Bryant, F. D. *J. Theoret. Biol.* **1968**, *21*, 348-367.
25. Deshpande, S.; Hall, J. C. *Biochim. Biophys. Acta* **1995**, *1244*, 69-78.
26. Deshpande, S.; Hall, J. C. *Pestic. Biochem. Physiol.* **1996**, *56*, 26-43.
27. Wang, Y.; Deshpande, S.; Hall, J. C. *Weed Sci.* **2001**, 49, 2-7.
28. Jones, A.M.; Im, K.H.; Savka, M.A.; Wu, M.J; DeWitt, N.G.; Shillito, R.; Binns, A.N. *Science* **1998,** *282*, 1114-1117.
29. Webb, S. R.; Hall, J. C. *Pestic. Biochem. Physiol.* **1995**, *52*, 137-148.

30. Deshpande, S.; Hall, J. C. *Pestic. Biochem. Physiol.* **1999**, *65*, in press.
31. Estelle, M. A.; Sommerville, C. *Mol. Gen. Genet.* **1987**, *206*, 200-206.
32. Leyser, H. M. O.; Lincoln, C. A.; Timpte, C. S.; Lammer, D.; Turner, J.C.; Estelle, J.C. *Nature* **1993**, *304*, 161-164.
33. Leyser, H. M. O. *Current Biology* **1998**, *8*, R305-R307.
34. Lincoln, C.; Britton, J. H.; Estelle, M. *Plant Cell* **1990**, *2*, 1071-1077.

Chapter 8

Cases and Mechanisms of Resistance to ACCase-Inhibiting Herbicides

K. K. Hatzios

Department of Plant Pathology, Physiology, and Weed Science, Virginia Polytechnic Institute and State University, Blacksburg, VA 24061–0330

Aryloxyphenoxypropionate (AOPP) and cyclohexanedione (CHD) herbicides have been used widely for the control of many grass weeds since their respective introductions in the 1970s and 1980s. The target site of AOPP and CHD herbicides is the biotin-containing enzyme acetyl-coenzyme A carboxylase (EC 6.4.1.2, ACCase), which catalyzes the ATP-dependent conversion of acetyl-coenzyme A to malonyl-coenzyme A in the fatty acid biosynthetic pathway. Grass plants are sensitive to these herbicides because they contain only the herbicide-sensitive eukaryotic form of ACCase in their plastids and cytosol. Because of the frequent and widespread use of these graminicides, fourteen grass weed species have developed resistance to AOPP and CHD herbicides in several countries around the world. The predominant mechanism of weed resistance to ACCase inhibiting herbicides appears to involve alterations of the target ACCase enzyme. Alternative mechanisms such as overexpression of ACCase activity, enhanced herbicide metabolism, and reduced effects of the AOPP and CHD herbicides on the electrogenic potential of plasma membrane have been proposed to explain the development of resistance to AOPP and CHD herbicides in selected grass weeds. The mechanisms of weed resistance to ACCase inhibiting herbicides and the case of an accession of smooth crabgrass [*Digitaria ischaemum* (Schleb.) Muhl.], which has developed resistance to the herbicide fenoxaprop-ethyl are discussed in this chapter.

Following their respective introduction in the 1970s and 1980s, aryloxyphenoxypropionate (AOPP) and cyclohexanedione (CHD) herbicides have been used widely for the control of many grass weeds (*1-4*). AOPP and CHD

© 2002 American Chemical Society

herbicides are postemergence herbicides used for the control of annual and perennial grasses in crops such as soybean, turf and winter cereals (*1-4*). The chemical structures of representative members of the AOPP and CHD herbicide classes are shown in Figure 1.

Inhibition of Acetyl-CoA Carboxylase by AOPP and CHD Herbicides

The target site of AOPP herbicides such as diclofop, fenoxaprop and quizalofop and CHD herbicides such as sethoxydim and clethodim is the biotin-containing enzyme acetyl-coenzyme A carboxylase (EC 6.4.1.2; ACCase) (*3-8*). ACCase is the first committed enzyme in the *de novo* fatty acid biosynthetic pathway. This enzyme was discovered in 1959 and is present in mammalian, prokaryotic and plant systems (*9-14*). ACCase catalyzes the ATP-dependent conversion of acetyl-coenzyme A to malonyl-coenzyme A. In plants, ACCase is a plastidic and/or cytosolic enzyme, which consists of three reaction regions: a biotin carboxylase which catalyzes the ATP-dependent carboxylation of biotin with HCO_3^-; a transcarboxylase which transfers the activated carbon dioxide from the biotin to acetyl-CoA; and the biotin-carrier peptide (BCP) which is covalently linked to the biotin group and allows the movement of biotin between the two catalytic centers (*9-14*). The ACCAse reactions can be summarized by the following equations:

$$ATP + HCO_3^- + BCP \xrightarrow{Mg^{2+}} CO_2\text{-}BCP + ADP + Pi \qquad (1)$$

Biotin carboxylase reaction

$$CO_2\text{-}BCP + Acetyl\text{-}CoA \rightarrow BCP + malonyl\text{-}CoA \qquad (2)$$

Transcarboxylase reaction

ACCases are complex proteins, whose subunit composition is not fully characterized (*9-14*). Plant ACCase have been proposed to be: a) homodimers of polypeptide subunits (219 or 227-kDa); b) multimers of high molecular weight composed of nonidentical polypeptides; or c) combinations of the previous two isoforms (12). Studies on cloning, sequencing, and regulation of the expression of ACCase genes in plants are rather limited at the present time (*15-17*).

The establishment of ACCase as the target site for all AOPP and CHD herbicides has been supported by the work of several investigators (*18-21*) and the subject has been reviewed recently by Incledon and Hall (*12*). Dicotyledonous plants are resistant to these herbicides because they contain two forms of ACCase in their cells: a prokaryotic form in the plastids, which is insensitive to AOPP and CHD herbicides, and a herbicide-sensitive eukaryotic form in the cytosol (*22-24*). Grass plants are sensitive to these herbicides because they contain only the herbicide-sensitive eukaryotic form of ACCase in their plastids and cytosol (*22-24*).

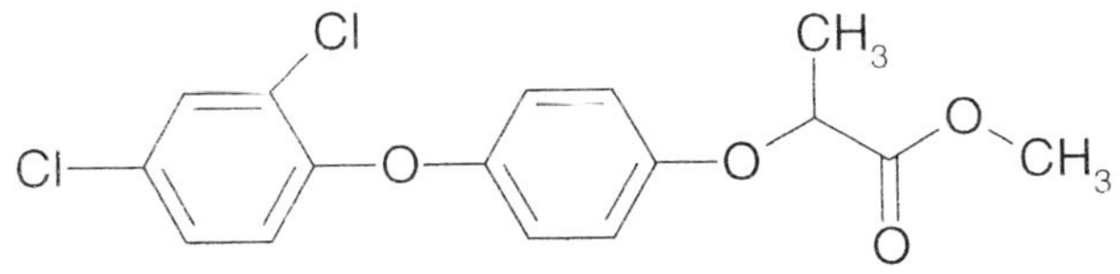

Diclofop-methyl

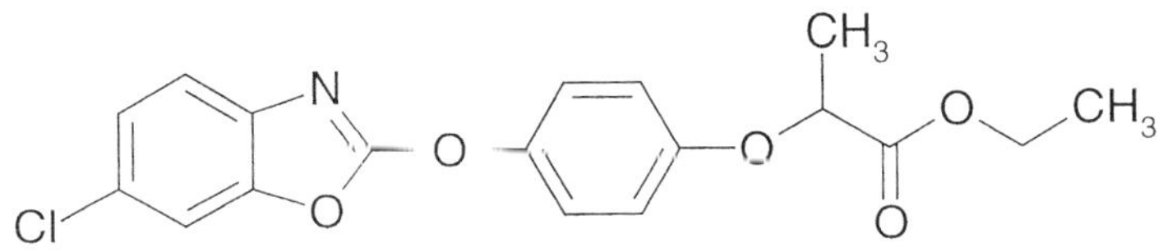

Fenoxaprop-P ethyl

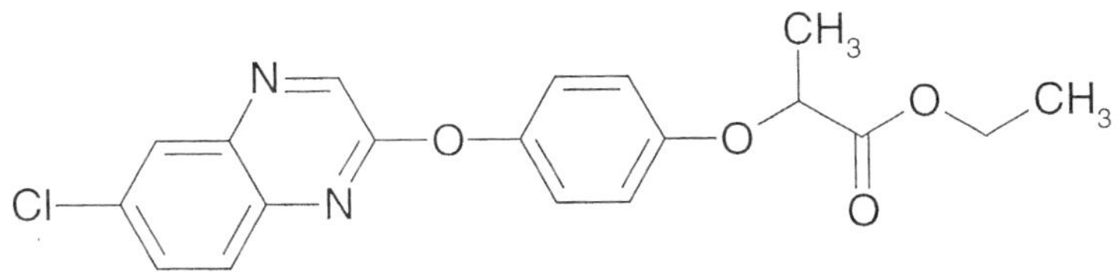

Quizalfop-P ethyl

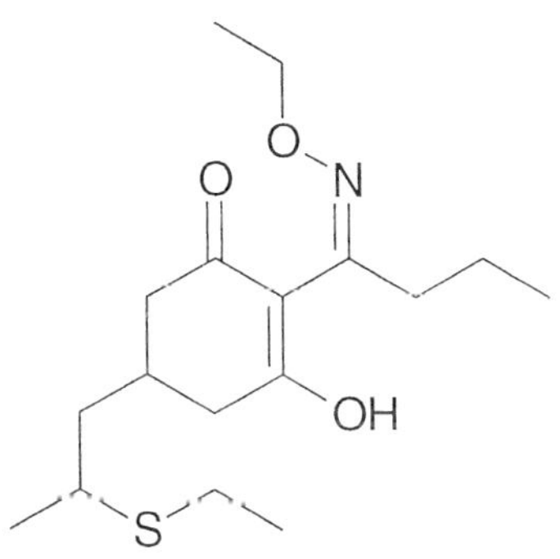

Sethoxydim

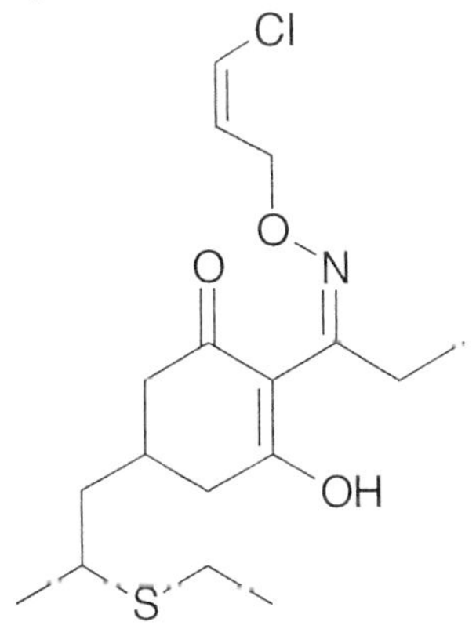

Clethodim

Figure 1. Chemical structures of the aryloxyphenoxypropionate (AOPP) herbicides diclofop-methyl, fenoxaprop-ethyl and quizalofop-ethyl and of the cyclohexanedione (CHD) herbicides sethoxydim and clethodim.

Development of Weed Resistance to ACCase-inhibiting Herbicides

The frequent and widespread use of the AOPP and CHD herbicides has resulted in the development of resistance in biotypes or accessions of several grass weeds, and the subject has been reviewed (*12, 25-27*). Resistant weeds confirmed outside the United States include annual ryegrass (*Lolium rigidum* Gaud), wild oat (*Avena fatua* L.) and winter wild oat (*Avena sterilis ssp. Ludoviciana* Malzew.) in Australia (*28-31*); wild oat and green foxtail [*Setaria viridis* (L.) Beauv.] in Canada (32-*34*); slender foxtail (*Alopecurus myosuroides* Huds.) in England (*35*) and Spain (*36*); goosegrass [*Eleusine indica* (L.) Gaertn.] in Malaysia (*37*); barnyardgrass [*Echinochloa colona* (L.) Link] in Costa Rica (27), and littleseed canarygrass (*Phalaris minor* Retz.) in Israel (*38*).

The occurrence of grass weed resistance to AOPP and CHD herbicides has been confirmed also in the United States. Diclofop-resistant biotypes of Italian ryegrass (*Lolium multiflorum* Lam.) and of wild oat have been confirmed in Oregon (*39, 40*) and wild oat in the Red River Valley (*41*). Two biotypes of Johnsongrass [*Sorghum halepense* (L.) Pers.] resistant to fluazifop and quizalofop have been confirmed in Mississippi (*42*), whereas a Johnsongrass biotype resistant to quizalofop and sethoxydim has been reported in Virginia (*43*). Accessions of large crabgrass [*Digitaria sanguinalis* (L.) Scop.] and giant foxtail (*Setaria faberi* Herrm.) exhibiting resistance to fluazifop-butyl and sethoxydim has been confirmed in Wisconsin (*44, 45*). Biotypes of red fescue (*Festuca rubra* L.) with resistance to sethoxydim have been confirmed in mid-western states (*46*). Finally, a smooth crabgrass accession resistant to fenoxaprop-ethyl was discovered recently in New Jersey (*47, 48*). Table 1 summarizes the weed species that are known to have developed resistance to ACCase-inhibiting herbicides around the world.

Mechanisms of Resistance to ACCase Inhibitors

Resistance Based on Alterations of ACCase

The resistance of grass weeds to ACCase-inhibiting herbicides appears to result mainly from alterations of the target ACCase enzyme (*12, 25, 27*). Studies on the mechanism of resistance have confirmed that reduced sensitivity of ACCase to AOPP and/or CHD herbicides is responsible for the development of resistant biotypes or accessions in the following grassy weeds: littleseed canarygrass to fenoxaprop (*38*); red fescue to sethoxydim (*46*); Italian ryegrass to diclofop (*49*); green foxtail to sethoxydim (*50, 51*), wild oats to sethoxydim (*27*) and diclofop-methyl (*51-53*); winter wild oat to fenoxaprop (*54*); large crabgrass to fenoxaprop (*45*); giant foxtail to fluazifop-butyl and sethoxydim (*51*); goosegrass to fluazifop, fenoxaprop, and sethoxydim (*56*), Johnsongrass to fluazifop (*57*); slender foxtail to fluazifop and fenoxaprop (*61*); and ryegrass to diclofop and fluazifop (*58, 59*).

Table I. Weed Species Exhibiting Resistance to ACCase-Inhibiting Herbicides

Species	*Location*	*Reference*
Avena fatua	Canada, USA, Australia, UK	*27, 30-32*
Avena sterilis	Australia, UK	*27, 30, 55*
Alopecurus myosuroides	UK, Spain, Germany, France	*27, 36*
Digitaria ischaemum	USA (New Jersey)	*47, 48*
Digitaria sanguinalis	USA (Wisconsin)	*45*
Echinochloa colona	Costa Rica	*27*
Eleusine indica	Malaysia	*37*
Festuca rubra	USA	*46*
Lolium rigidum	Australia, Spain	*27-30*
Lolium multiflorum	UK, USA (Oregon)	*39*
Phalaris minor	Israel	*38*
Setaria faberi	USA (Iowa, Wisconsin)	*34, 51*
Setaria viridis	Western Canada	*44*
Sorghum halepense	USA (Mississippi, Virginia)	*42, 43*

NOTE: Resistance has been conferred to one or more biotypes or accessions of the above weed species by reduced ACCase sensitivity.

Other Mechanisms of Resistance

While the evidence supporting alterations of ACCase as the main mechanisms of weed resistance to AOPP and CHD herbicides, alternative mechanisms of resistance have been also proposed (*27, 60*). Overexpression of ACCase activity appears to be related to the development of Johnsongrass resistance to sethoxydim (*43*). Reduced absorption and translocation have not been implicated as mechanisms for the resistance of grass weeds to AOPP and CHD herbicides (*27, 60*). However, enhanced herbicide metabolism has been suggested as a mechanism of resistance to ACCase inhibitors in selected biotypes of slender foxtail (*36*), annual ryegrass (*64*) and large crabgrass (*27*). Finally, although a specific mechanism of resistance was not identified, evidence-linking resistance to a differential effect of AOPP and CHD herbicides on the electrogenic potential of plasma membranes has been presented in studies with slender foxtail and ryegrass (*62-64*).

Patterns of Weed Cross-resistance to ACCase-inhibiting Herbicides

Although AOPP and CHD herbicides bind in the same region of the target ACCase enzyme (*12, 21*), different cross-resistance patterns appear to characterize the resistance of grass weeds to ACCase inhibitors (*65*). In general, three different types of weed cross-resistance to AOPP and CHD herbicides have been characterized (*65*). Cross-resistance type C is the most common and is characterized by high levels of resistance to both AOPP and CHD herbicides (*65*). Cross-resistance type A is highly resistant to the AOPP herbicides, but shows little or no resistance to CHD herbicides (*65*). Cross-resistance type B shows low or moderate resistance to both groups. Weeds showing cross-resistance type A include wild oats, Italian ryegrass, and smooth crabgrass biotypes and/or accessions, which have high resistance to AOPP herbicides, but almost no resistance to CHD herbicides (*27, 47, 48, 65*). Cross-resistance patterns in green foxtail (*50*), large crabgrass (*45*), giant foxtail (*44*), goosegrass (*37*), and Johnsongrass (*42*) resemble the patterns that characterize types B and C. A fourth pattern of resistance includes biotypes of the weeds green foxtail and giant foxtail which are resistant to sethoxydim, but had only low level of resistance to other AOPP and CHD herbicides (51). The basis of the differential cross-resistance types may be related to specific point mutations that alter the ACCase binding sites in different ways.

Alternative patterns of cross-resistance have been also reported for biotypes and/or accessions of other grass weeds. Thus, a wild oat biotype showed very high (150-fold) resistance to sethoxydim, but much lower resistance to other AOPP and CHD herbicides (*32, 53*). Similarly, a biotype of giant foxtail from Wisconsin was highly (>145-fold) resistant to sethoxydim and resistant (25-fold) to fluazifop-butyl in greenhouse experiments (*44*). A commercialized sethoxydim-resistant line of corn is cross-resistant (16-fold) to haloxyfop at the ACCase level (*66*).

Other resistant weed biotypes appear to result from different ACCase mutations, conferring quite different patterns of resistance to the same herbicide. For example, ACCase from a resistant goosegrass biotype was very resistant to fluazifop but much less resistant to sethoxydim, fenoxaprop, and clethodim (*56*). Similarly, ACCase from a resistant biotype of annual ryegrass was very resistant to diclofop, but much less resistant to CHD herbicides (*67*).

Case Study: Smooth Crabgrass Resistance to Fenoxaprop-ethyl

An accession of smooth crabgrass [*Digitaria ischaemum* (Schreb.) Muhl] exhibiting resistance to the herbicide fenoxaprop was discovered recently in New Jersey (*47*). Kuk *et al.* (*48*) showed that this accession was highly resistant to fenoxaprop-ethyl and moderately resistant to quizalofop-ethyl, but exhibited low or no cross-resistance to CHD herbicides such as sethoxydim.

Growth Responses

Seedlings of resistant and susceptible accessions of smooth crabgrass were grown from seed and maintained in a greenhouse with 25°C temperature, 16-h photoperiod, and natural plus artificial lighting providing a photon flux density of 650 μmol/m^2/s. When seedlings reached the two-tiller stage they were treated with commercial formulations of fenoxaprop-ethyl, quizalofop-P-ethyl, and sethoxydim using a belt-link sprayer equipped with a single flat fan nozzle (8001E) delivering 233.7 L/ha. The recommended field rate of fenoxaprop-ethyl is 0.15 kg/ha. Treatment rates used were 0, 0.03, 0.05, 0.07, 0.14, 0.28, 0.56, 1.12, and 2.24 kg/ha for fenoxaprop-ethyl; 0, 0.01, 0.03, 0.07, 0.14, 0.28, 0.56, and 0.84 kg/ha for quizalofop-P-ethyl; and 0, 0.05, 0.1, 0.2, and 0.4 kg/ha for sethoxydim. Treated plants were returned to the greenhouse. Two weeks after treatment, seedlings of the resistant and susceptible accessions of smooth crabgrass were evaluated for their growth responses to fenoxaprop-ethyl, quizalofop-ethyl, and sethoxydim.

Figures 2, 3, and 4 show the growth responses of two smooth crabgrass accessions following treatment with two AOPP and one CHD herbicides. The resistant accession of smooth crabgrass exhibits high resistance to fenoxaprop-ethyl (Figure 2), good cross-resistance to quizalofop-ethyl (Figure 3) and no cross-resistance to sethoxydim (Figure. 4). The calculated GR_{50} dose of fenoxaprop-ethyl on susceptible smooth crabgrass was 0.11 kg/ha compared to 11.2 kg/ha for the resistant accession (*48*). Thus, the resistant smooth crabgrass accession had greater than 102-fold resistance to fenoxaprop-ethyl relative to the susceptible smooth crabgrass. The calculated GR_{50} values of quizalofop-ethyl were 0.65 kg/ha on the resistant accession and 0.04 kg/ha for the susceptible smooth crabgrass (*48*). The calculated GR_{50} values of sethoxydim were very similar in the resistant and susceptible accession of smooth crabgrass (0.08 vs. 0.06 kg/ha).

The morphology and growth of resistant smooth crabgrass seedlings were comparable to those of susceptible smooth crabgrass seedlings when grown in the greenhouse. Wiederholt and Stoltenberg (*68, 69*) have reported that the development of resistance to AOPP and CHD herbicides in accessions of large crabgrass and giant foxtail was not associated with reduced fitness.

Mechanism of Resistance

Absorption, Translocation, and Metabolism

The rate of absorption of radioactivity following exposure to leaf-applied radiolabeled fenoxaprop-ethyl was similar between the two accessions of smooth crabgrass (*48*). The majority of the absorbed ^{14}C remained in the treated leaf at all sampling times indicating that fenoxaprop did not translocate appreciably out of the treated leaf in both accessions of smooth crabgrass (*48*). These results suggest that

Figure 2. Dose-response effects of fenoxaprop-ethyl on the growth of resistant (top) and susceptible (bottom) smooth crabgrass two weeks after treatment with 0, 0.03, 0.05, 0.07, 0.14, 0.28, 0.56, 1.12, and 2.24 kg/ha.

Figure 3. Dose-response effects of quizalofop-ethyl on growth of resistant (top row) and susceptible (bottom row) smooth crabgrass two weeks after treatment with 0, 0.01, 0.03, 0.07, 0.14, 0.28, 0.56, and 0.84 kg/ha.

Figure4. Dose-response effects of sethoxydim on the growth of resistant (top row) and susceptible (bottom row) smooth crabgrass two weeks after treatment with 0, 0.05, 0.1, 0.2, and 0.4 kg/ha.

differential absorption and translocation are not involved in the observed resistance of smooth crabgrass to the herbicide fenoxaprop.

Because of the lack of appreciable herbicide translocation, the metabolism of ^{14}C-fenoxaprop-ethyl was examined only in extracts from treated leaves of resistant and susceptible smooth crabgrass seedlings (*48*). The obtained results indicated that differential metabolism is not a factor contributing to the observed resistance of smooth crabgrass to the herbicide fenoxaprop (*48*). These results were consistent with previous reports demonstrating that the resistance of many grass weeds to AOPP and CHD herbicides is not conferred by altered herbicide absorption, translocation or metabolism (*12, 27*).

ACCase Sensitivity

ACCase extraction, purification, and assaying procedures were performed according to Mannechote et al. (55) with the modifications described by Kuk et al. (48). Fenoxaprop-ethyl was dissolved in acetone (0.1% final concentration) and all dilutions were made with Tricine-KOH buffer (pH 8.3) to yield final assay concentrations of 0.01, 1, 10, and 100 μM. The herbicidal concentration causing a 50% inhibition of ACCase activity (I_{50}) was calculated from the concentration-response curve shown in Figure 5.

The specific activity of ACCase extracted from shoot tissues of susceptible and resistant smooth crabgrass was similar (*48*). Data in Figure 5 show that ACCase extracted from susceptible plants was 91 times more sensitive to inhibition by fenoxaprop-ethyl (I_{50} of 2 μM) than ACCase extracted from resistant smooth crabgrass (I_{50} of 182 μM). The activity of ACCase extracted from resistant plants was also less sensitive to fenoxaprop, quizalofop-P-ethyl and sethoxydim than that extracted from susceptible plants (*48*). The respective R/S I_{50} ratios were 5.9 for fenoxaprop, 41.3 for quizalofop, and 2.2 for sethoxydim (*48*).

These data suggest that a less sensitive form of acetyl-CoA carboxylase confers fenoxaprop resistance to the accession of smooth crabgrass discovered in New Jersey. The basis of fenoxaprop resistance in smooth crabgrass appears to be similar to that of biotypes and/or accessions of green foxtail, annual ryegrass, Italian ryegrass, and winter wild oat, whose resistance to AOPP and CHD herbicides has been linked to reduced sensitivity of the target enzyme. A single dominant or partially dominant nuclear gene (*66, 67*) controls the target enzyme based resistance to ACCase inhibitors. This suggests that resistance can be conferred by a number of different point mutations, each of which confers a different pattern of resistance and cross-resistance to AOPP and CHD herbicides.

Concluding Remarks

The development of resistance to ACCase-inhibiting herbicides in several grass weeds has been confirmed and is an increasing problem in several parts of the world.

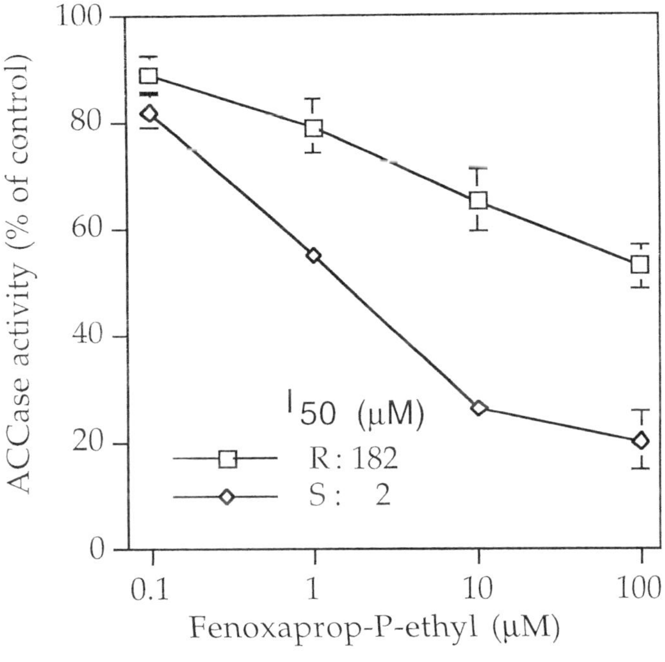

Figure 5. Inhibition of ACCase activity extracted from shoot tissues of susceptible (S) and resistant (R) smooth crabgrass following treatment with fenoxaprop-ethyl. Vertical bars represent standard errors of means (n = 9) and they are not shown when less than the marker size.

Resistance to these herbicides can arise easily following selection pressure with AOPP and CHD herbicides for six to ten years (*27*). Although alternative mechanisms have been proposed, the predominant mechanism of grass weed resistance to these herbicides results from the reduced sensitivity of ACCase in resistant mutants. Future advances in sequencing of mutated ACCases from resistant weeds are expected to provide us with a better understanding of the molecular mechanism of resistance. Multiple target-site mutations have developed in some of the resistant weed biotypes, severely limiting weed control options with chemical herbicides. The judicious use of ACCase inhibiting herbicides in combination with herbicides from other classes and methods of non-chemical weed control will be important for prolonging the usefulness of the AOPP and CHD herbicides. Improved sanitation of tillage and harvesting equipment and the use of certified crop seed has been reported as an important strategy for limiting the seed movement of weeds that have developed resistance to ACCase inhibitors (*70*). A seed bioassay, amplified fragment length polymorphism (AFLP) analysis, and mapping of risk areas have been used successfully for monitoring the spread of biotypes and accessions of wild oats that have developed resistance to ACCase inhibitors (*70-73*). Knowledge of the inheritance of resistance may help in the development of containment measures to prevent the spread of resistant genes from weeds that are resistant to ACCase inhibitors (*74*). The recent commercialization of transgenic crops such as soybeans, which are resistant to glyphosate, glufosinate, and selected sulfonylurea herbicides, provides growers with additional management options for controlling Johnsongrass and other grass weeds that have developed resistance to AOPP and CHD herbicides

Acknowledgments

The author would like to thank Drs. Yong-In Kuk, Jingrui Wu, and Jeff Derr for past collaboration and their contributions in the previously published study on the mechanism of resistance of smooth crabgrass to fenoxaprop-ethyl.

References

1. Duke, S. O.; Kenyon, W.H. In *Herbicides - Chemistry, Degradation, and Mode of Action,* Kearney, P. C.; Kaufman, D.D., Eds.; Marcel Dekker: New York, NY, 1988; Vol. 3, pp. 71-116.
2. Ishikawa, H.; Yamada, S.; Hosaka, H.; Kawana, T.; Okukuki, S.; Kohara, K. *J. Pestic. Sci.* **1985**, *10*, 187.
3. Walker, K.A.; Ridley, S.M.; Lewis, T.; Harwood, J. L. *Rev. Weed Sci.* **1989**, *4*, 71.
4. Gronwald, J. W. *Weed Sci.* **1991**, *39*, 435.
5. Gronwald, J. W. *Biochem. Soc. Transactions.* **1994**, *22*, 616.

6. Lichtenthaler, H. K. *Z. Naturforsch.* **1990**, *45c,* 521.
7. Goltz, A.; Focke, M.; Lichtenhaler, H. K. *J. Plant Physiol.* **1994**, *143*, 452.
8. Harwood, J. L.; Ridley, S. M.; Walker, K. A. *Soc. Exp. Biol. Seminar Series.* **1989**, *38*, 73.
9. Hellyer, A.; Bambridge, H. E.; Slabas, A. R. *Biochem. Soc. Trans.* **1986**, *14*, 565.
10. Harwood, J. L. *Proc. Brighton Crop Prot. Conf., Weeds* **1989**, *3B-2*, 155.
11. Harwood, J. L. *Annu. Rev. Plant Physiol. Plant Mol. Biol.* **1988**, *39*, 101.
12. Incledon, B. J.; Hall, J. C. *Pestic. Biochem. Physiol.* **1997**, *57*, 255.
13. Gornicki, P; Haselkorn, R. *Plant Mol. Biol.* **1993**, *22*, 547.
14. Egli, M. A.; Gengenbach, B. G.; Gronwald, J. W.; Somers, D. A.; Wyse, D. L. *Plant Physiol.* **1993**, *101*, 499.
15. Somers, D. A.; Keith, R. A.; Egli, M. A.; Marshall, L. C.; Gengenbach, B. G.; Grownald, J. W.; Wyse, D. L. *Plant Physiol.* **1993**, *101*, 1097.
16. Ashton, A. R.; Jenkins, C. L. D.; Whitfield, P. R. *Plant Mol. Biol.* **1994**, *24*, 35.
17. Shorrosh, B. S.; Roesler, K. R.; Shintani, D.; van de Loo, F.J.; Ohlrogge, J. B. *Plant Physiol.* **1995**, *108*, 805.
18. Burton, J. D.; Gronwald, J. W.; Somers, D. A.; Connelly, J. A.; Gengenbach, B. G.; Wyse, D. L. *Biochem. Biophys. Res. Comm.* **1987**, *148*, 1039.
19. Rendina, A. R.; Felts, J. M.; Beaudoin, J. D.; Craig-Kennard, A. C.; Look, L. L.; Paraskos, S. L.; Hagenah, J. A. *Arch. Biochem. Biophys.* **1988**, *265*, 219.
20. Secor, J.; Cseke, C.; Owen, W. J. *Proc. Brighton Crop Prot. Conf. Weeds* **1989**, *3B-1*, 145.
21. Rendina, A. R.; Craig-Kennard, A. C.; Beaudoin, J. D.; Breen, M. K. *J. Agric. Food Chem.* **1990**, *38*, 1282.
22. Konishi, T.; Sasaki, Y. *Proc. Natl. Acad. Sci. USA.* **1994**, *91*, 6860.
23. Sasaki, Y; Konishi, T; Nakano, Y. *Plant Physiol.* **1995**, *108,* 445.
24. Konishi, T.; Shinohara, K.; Yamada, K; Sasaki, Y. *Plant Cell Physiol.* **1996**, *37*, 117.
25. Devine, M.D.; Shimabukuro, R. H. In: *Herbicide Resistance in Plants: Biology and Biochemistry*, Polwes, S.B.; Holtum, J. A. M., Eds.; Lewis Publishers, Boca Raton, FL. 1994; pp. 141-149.
26. Somers, D. A. In: *Herbicide Resistant Crops*; Duke, S. O., Ed.; CRC Press, Boca Raton, FL.; 1996; pp. 175-188.
27. Devine, M. D. *Pestic. Sci.* **1997**, *51*, 259.
28. Heap, I. M.; Knight, R. *J. Aust. Inst. Agric. Sci.* **1982**, *48*, 156.
29. Heap, I. M.; Knight, R. *Aust. J. Agric. Res.* **1990**, *41*, 121.
30. Powles, S. B.; Howat, P. D *Weed Technol.* **1990**, *4,* 178.
31. Mansooji, A. M.; Holtum, J. A.; Boutsalis, P; Mattews, J. M.; Powles, S. B. *Weed Sci.* **1992**, *40,* 599.
32. Heap, I. M.; Murray, B. G.; Loeppky, H. A.; Morrison, I. M. *Weed Sci.* **1993**, *41,* 232.
33. Joseph, O. O.; Hobbs, S. L. A.; Jana, S. *Weed Sci.* **1990**, *38,* 475.
34. Heap, I. M.; Morrison I. N. *Weed Sci.* Soc. Am. Abstr. **1993**, *33,* 62.
35. Moss, S. R. *Weed Sci.* **1990**, *38,* 492.

36. Menendez, J.; De Prado, R.; Jorrin, J.; Taberner, A. *Proc. Brighton Crop Prot. Conf. Weeds*, **1993**, *1,* 213.
37. Marshall, G.; Kirkwood, R. C.; Leach, G. E. *Weed Res.* **1994**, *34,* 177.
38. Tal, A.; Zarka, S.; Rubin, B. *Pestic. Biochem. Physiol.* **1996**, *56,* 134.
39. Stanger, A. E.; Appleby, A. P. *Weed Sci.* **1989**, *37,* 350.
40. Seefeldt, S. S.; Gealy, D. R.; Brewster, B. D.; Fuert, E. P. *Weed Sci.* **1994**, *42,* 430.
41. Frie, J. B.; Zollinger, R. K.; Manthey, F. A.; Durgan, R. R. *Proc. North Cent. Weed Sci. Soc.* **1993**, *48,* 16.
42. Smeda, R. J.; Snipes, C. E.; Barrentine, W. L . *Weed Sci.* **1997**, *45*, 132.
43. Bradley, K. W.; Wu, J.; Hagood, E. S.; Hatzios, K. K. *Abstr. Weed Scie. Soc. Am.* **1999**, *39*, 000 .
44. Stoltenberg, D. E.; Wiederholt, R. J. *Weed Sci.* **1995**, *43,* 527.
45. Wiederholt, R. J.; Stoltenberg, D. E. *Weed Technol.* **1995**, *9,* 518.
46. Stoltenberg, D. E.; Gronwald, J. W.; Wyse, D. L.; Burton, J. D.; Somers, D. A.; Gengenbach, B. G. *Weed Sci.* **1989**, *37*, 512.
47. Derr, J. F. *Proc. Northeast. Weed Sci. Soc.* **1998**, *52,* 85.
48. Kuk, Y-I.; Wu, J.; Derre, J. F.; Hatzios, K. K. *Pestic. Biochem. Physiol.* **1999**, *64*, 112.
49. Gronwald, J. W.; Eberlein, C. V.; Betts, K. J.; Baerg, R. J.; Ehlke, N. J.; Wyse, D. L. *Pestic. Biochem. Physiol.* **1992**, *44,* 126.
50. Marles, M. A. S.; Devine, M. D.; Hall, J. C. *Pestic. Biochem. Physiol.* **1993**, *46,* 7.
51. Shukla, A.; Leach, G. E.; Devine, M. D. *Plant Physiol. Biochem.* **1997**, *35*, 803.
52. Devine, M. D.; MacIsaac, S. A.; Romano, M. L.; Hall. J. C. *Pestic. Biochem. Physiol.* **1992**, *42,* 88.
53. Shukla, A.; Dupont, S.; Devine, M. D. *Pestic. Biochem. Physiol.* **1997**, *57,* 147.
54. Seefeldt, S. S.; Fuerst, E. P.; Gealy, D. R.; Shukla, A.; Irzyk, G. P.; Devine, M. D. *Weed Sci.* **1996**, *44,* 776.
55. Maneechote, B; Holtum, J. A. M.; Preston, C.; Powles, S. B. *Plant Cell Physiol.* **1994**, *35,* 627.
56. Leach, A. E.; Devine, M. D.; Kirkwood, R. C.; Marshall, G. *Pestic. Biochem. Physiol.* **1995**, *51,* 129.
57. Obermeier, M. R.; Barrett, M.; Witt, W. W.; Green, J. D. *Abstr. Weed Sci. Soc. Amer.* **1997**, *37*, 64.
58. Tardif, D. J.; Holtum, J. A. M.; Powles, S. B. *Planta* **1993**, *190,* 176.
59. Preston, C.; Tardif, F. J.; Christopher, J. T.; Powles, S. B. *Pestic. Biochem. Physiol.* **1996**, *54,* 123.
60. Devine, M. D.; Renault, S.; Wang, X. *Proc. Brighton Crop Prot. Conf. Weeds,* **1993**, *2*, 541.
61. Hall, L. M.; Moss, S. R.; Powles, S. B. *Pestic. Biochem. Physiol.* **1997**, *xx*, 000.
62. Shimabukuro, R. H.; Hoffer, B. L. *Plant Physiol.* **1992**, *98,* 1415.
63. Devine, M. D.; Hall, J. C.; Romano, M. L.; Marles, M. A. S.; Thomson, L. W.; Shimabukuro, R. H. *Pestic. Biochem. Physiol.* **1993**, *45,* 167.
64. Hausler, R. E.; Holtum, J. A. M.; Powles, S. B. *Plant Physiol.* **1991**, *97*, 1035.

65. Bourgeois, L.; Kenkel, N. C.; Morrison, I. N. *Weed Sci.* **1997**, *45*, 750.
66. Parker, W. B.; Marshall, L. C.; Burton, J. D.; Somers, D. A.; Wyse, D. L.; Gronwald, J. W.; Gengenbach, B. G. *Proc. Natl. Acad. Sci. USA*. **1990**, *87,* 7175.
67. Tardif, F. J.; Preston, C.; Holtum, J. A. M.; Powles, S. B. *Aust. J. Plant Physiol.* **1996**, *23,* 15.
68. Wiederholt R. J.; Stoltenberg, D. E. *Weed Technol.* **1996**, *10,* 42.
69. Wiederholt, R. J.; Stoltenberg, D. E. *Weed Sci.* **1996**, *44,* 18.
70. Andrews, T. S.; Morrison, I. N.; Penner, G. A. *Weed Sci.* **1998**, *46*, 196.
71. Bourgeois, L.; Morrison, I. N. *Can. J. Plant Sci.* **1997**, *77*, 173.
72. Bourgeois, L.; Morrison, I. N. *Can. J. Plant Sci.* **1997**, *77*, 703.
73. Murray, B. G.; Brule-Babel, A. L.; Morrison, I. N. *Weed Sci.* **1996**, *44,* 476.
74. Seefeldt, S. S.; Hofffman, D. L.; Gealy, D. R.; Fuerst, E. P. *Weed Sci.* **1998**, *46*, 170.

Chapter 9

Mechanisms of Multiple Herbicide Resistance in *Lolium rigidum*

Christopher Preston[1] and Stephen B. Powles[2]

[1]Department of Applied and Molecular Ecology and CRC for Weed Management Systems, Waite Campus, University of Adelaide, PMB 1 Glen Osmond, SA 5064, Australia
[2]Western Australian Herbicide Resistance Initiative, Faculty of Agriculture, University of Western Australia, Nedlands, WA 6907, Australia

Herbicide-resistant populations of *Lolium rigidum* infest more than 8,000 farms in southern Australia. These populations may be resistant to one or more herbicides from many modes of action including inhibitors of lipid biosynthesis, branched-chain amino acid biosynthesis, photosynthesis, tubulin polymerization, carotenoid biosynthesis, or aromatic amino acid biosynthesis. The mechanisms of resistance have been extensively studied, and populations with herbicide-insensitive acetyl-coenzyme A carboxylase and acetolactate synthase enzymes are known. In addition, many populations have the ability to rapidly metabolize a number of herbicides from several different modes of action. In contrast, the mechanism of resistance to glyphosate is due to neither an insensitive target site, nor to more rapid glyphosate metabolism. *L. rigidum* populations readily accumulate multiple resistance mechanisms and this creates major difficulties in managing this ubiquitous grass weed.

© 2002 American Chemical Society

Introduction

Lolium rigidum is a widespread grass weed present across the entire grain-growing area of southern Australia. Populations of *L. rigidum* can easily exceed 2,000 plants m^{-2} in crops making it important to control this weed (*1*). The extensive and intensive use of herbicides to control *L. rigidum* populations in crops has resulted in the evolution of herbicide resistance. The first reported case of herbicide resistance was in 1982 for a population resistant to diclofop-methyl (*2*). Since then, numerous other populations have evolved herbicide resistance. Currently, more than 8,000 grain growers report herbicide resistant *L. rigidum* populations (*3*).

Two major factors have combined to make herbicide resistance in *L. rigidum* a major management problem for grain growers. The first is the speed with which herbicide resistance evolves in populations of this species. For example, resistance to the acetyl-coenzyme A carboxylase and acetolactate synthase inhibitors can occur with three or four applications of these herbicides (*4, 5*). The second factor is the propensity with which populations evolve multiple resistance to a wide range of chemicals (*6, 7*). In the worst case, a population can be simultaneously resistant to herbicides from more than ten different chemical groups inhibiting six different target sites (*7*).

The combination of a huge number of herbicide-resistant weed populations and the presence of multiple herbicide resistance in these populations have placed tremendous pressure on weed control in grain crops in Australia. Changes to cropping practices resulting from the widespread evolution of resistance to the selective post-emergent herbicides have meant greatly increased reliance on pre-seeding herbicides such as trifluralin and glyphosate. This has been followed in recent years by an increased number of reports of trifluralin resistance (Preston, C. unpublished data) and by the evolution of glyphosate resistance (*8, 9*) in *L. rigidum* populations.

L. rigidum populations have evolved resistance due to resistant target enzymes, enhanced herbicide metabolism, and other mechanisms. These different resistance mechanisms are discussed below.

Resistance Due to Insensitive Target Enzymes

Resistance to Acetolactate Synthase (ALS) Inhibitors

The sulfonylurea herbicides, chlorsulfuron and triasulfuron, were widely used for the control of *L. rigidum* across southern Australia. The result is the evolution of a large number of *L. rigidum* populations in Australia with resistance to inhibitors of ALS (*4*). Resistance to the sulfonylurea herbicides appeared rapidly and some populations show cross-resistance to imidazolinone herbicides (*4*).

Extraction and assay of ALS *in vitro* from some resistant populations of *L. rigidum* demonstrates that resistance can result from possession of a resistant target enzyme (*10*). Within these resistant ALS enzymes, variation is evident in the patterns of cross-resistance to herbicides from different chemistries of ALS inhibitors (Table I). Some populations are highly resistant to herbicides from the sulfonylurea, imidazolinone, and triazolopyrimidine sulfonamide chemistries, whereas others have different patterns of resistance.

Table I. Variation in Target Site Cross-Resistance to Chlorsulfuron, Flumetasulam and Imazapyr of ALS Isolated from Different Resistant *L. rigidum* Populations and Assayed *in vitro*.

Population	*Resistance Index[a] for Herbicide*		
	Chlorsulfuron	*Flumetsulam*	*Imazapyr*
WLR 1	783	103	30
SLR 89	267	23	14
SLR 112	283	24	29

[a] Resistance index is calculated as the concentration of herbicide required to inhibit the resistant enzyme by 50% divided by the concentration of herbicide required to inhibit the susceptible enzyme by 50%.

SOURCE: (Preston, C., unpublished data)

The ALS gene has been sequenced from many plant species and the locations of mutations that endow resistance to herbicides determined. In plants, amino acid substitutions at one of five locations within highly conserved parts of the ALS protein endow resistance (*11*). It is also known that different amino acid substitutions within ALS result in enzymes with different patterns of cross-resistance. For example, ALS enzymes might be resistant to both sulfonylurea and imidazolinone herbicides, to imidazolinone herbicides only, or to sulfonylurea herbicides only (*12, 13*). While the mutations endowing resistance in *L. rigidum* have yet to be determined, the enzyme inhibition data in Table I indicate that a variety of amino acid substitutions are likely. This data also demonstrates that more subtle variations in cross-resistance are possible. ALS isolated from all three populations is highly resistant to the sulfonylurea herbicide chlorsulfuron. ALS from WLR 1 is highly resistant to the triazolopyrimidine herbicide flumetsulam, whereas ALS from the other two resistant populations is not as resistant to this herbicide. Resistance to the imidazolinone herbicide imazapyr is less pronounced for all three populations; however, ALS from WLR 1 and SLR 112 have higher, and equal, resistance to this herbicide compared to ALS from SLR 89. These results suggest that three different amino acid substitutions are probably present in ALS from these three populations.

Resistance to ALS-inhibiting herbicides occurs very rapidly in *L. rigidum*, often within 4 years of use (*4*). The frequency of individuals with resistance to sulfonylurea herbicides in pristine *L. rigidum* populations that have never previously been exposed to herbicides is high (Table II). Frequencies of individuals with a resistant ALS vary from 1.9 x 10^{-5} to 1.1 x 10^{-4} in these populations. This high initial frequency of herbicide-resistant individuals in previously untreated weed populations means that resistance will evolve rapidly once the herbicides are used.

Table II. Initial Frequency of Sulfometuron-Methyl-Resistant Individuals in Three Pristine *L. rigidum* Populations.

Population	*Individuals Treated*	*Individuals Resistant*	*Frequency of Resistant Individuals*
VLR 1	202,000	12	5.9 x 10^{-5}
SLR 4	157,000	16	1.1 x 10^{-4}
SLR 15	155,000	3	1.9 x 10^{-5}

SOURCE: Preston, C. and Powles, S.B., unpublished data

Resistance to Acetyl-Coenzyme A Carboxylase (ACCase)-Inhibiting Herbicides

Resistance to ACCase-inhibiting herbicides in *L. rigidum* is also widespread across southern Australia (*4*). Resistance to the ACCase inhibitors occurs less rapidly than resistance to the ALS inhibitors; however, ACCase inhibitors have been used for longer. A range of ACCase inhibitors from two different herbicide chemistries is used to manage *L. rigidum* in Australia. This means that different resistant populations may have different selection histories. This is reflected in the wide variation in patterns of cross-resistance observed with different herbicide-resistant *L. rigidum* populations (Table III).

The ACCase isolated from herbicide-resistant populations of *L. rigidum* may be only resistant to aryloxyphenoxypropanoate herbicides, such as diclofop acid and fluazifop acid, or be resistant to cyclohexanedione herbicides as well. To date, no populations that are just resistant to the cyclohexanedione herbicides are known in Australia. However, populations of grass weeds resistant to cyclohexanedione, but not to aryloxyphenoxypropanoate, herbicides are known for other species elsewhere in the world (*11*).

Within this general pattern, other variations are also evident (Table III). For example, ACCase from VLR 69 has substantially greater resistance to diclofop acid compared to fluazifop acid; whereas ACCase isolated from SLR 31B has the opposite pattern of resistance. These variations in cross-resistance also occur with the cyclohexanedione herbicides. ACCase from SLR 74 is highly resistant to sethoxydim, but much less resistant to tralkoxydim; whereas, ACCase from WLR 33 has lower resistance to sethoxydim compared to tralkoxydim. It is highly likely that these variations in cross-resistance patterns are the result of different amino acid substitutions within the ACCase enzyme; however, at present the amino acid substitutions that endow herbicide resistance in ACCase are not known (*11*). The variations in cross-resistance patterns in *L. rigidum* populations have been exploited by grain growers in Australia who regularly use clethodim, and more recently butroxydim, to manage populations of *L. rigidum* resistant to other ACCase inhibitors.

Table III. Variations in Target Site Cross-Resistance to Aryloxyphenoxypropanoate and Cyclohexanedione Herbicides of ACCase Isolated from Resistant *L. rigidum* Populations and Assayed *in vitro*.

Population	*Resistance Index[a] for Herbicides*			
	Diclofop Acid	*Fluazifop Acid*	*Sethoxydim*	*Tralkoxydim*
SLR 3	>37	>>3	8	>10
SLR 31B	6	55	26	14
SLR 74	50	17	>50	7
VLR 69	29	4	1	1
WLR 33	368	>60	6	16
WLR 96	>217	>>7	>2	6

[a] Resistance index is calculated as the concentration of herbicide required to inhibit the resistant enzyme by 50% divided by the concentration of herbicide required to inhibit the susceptible enzyme by 50%.

SOURCE: Collated from (*5, 7, 14, 15*, Preston, C., unpublished data)

Resistance to Other Herbicides

In addition to resistance to the ACCase and ALS-inhibiting herbicides, populations of *L. rigidum* are resistant to photosystem II-inhibiting herbicides, trifluralin, glyphosate, and other herbicides. In contrast to the situation with resistant populations of other weed species, resistance to PS II inhibitors in *L. rigidum* is not due to alterations in the D1 protein of PS II (*16, 17*). Likewise, target site resistances to glyphosate (18) and trifluralin (Redden, E., Rieger, M., and Preston, C., unpublished data) have not been demonstrated in *L. rigidum* populations.

Resistance and Multiple Resistance Due to Increased Herbicide Detoxification

Populations of *L. rigidum* also evolve resistance to herbicides through increased herbicide detoxification. This phenomenon has been investigated in several resistant populations. These populations show varying patterns of resistance across a large number of herbicides (Table IV). In resistant plants, herbicides are detoxified at rates ranging from 1.4 to 4 times more rapidly than in susceptible plants, depending on the herbicide (*7, 10, 16, 17, 19, 20*). Where herbicide products have been identified in resistant plants, these have proven characteristic of cytochrome P450 monooxygenase-dependent reactions (*16, 17, 20, 21*).

Table IV. Variations in Metabolism-Based Herbicide Resistance in *L. rigidum* Populations.

Population	*Herbicides to Which Enhanced Metabolism is Demonstrated*
SLR 31	Diclofop-methyl
	Fluazifop-butyl
	Tralkoxydim
	Chlosulfuron
VLR 69	Simazine
	Chlorotoluron
	Metribuzin
	Diclofop-methyl
	Tralkoxydim
	Chlorsulfuron
WLR 1	Chlorsulfuron
WLR 2	Simazine
	Chlorotoluron
	Metribuzin

SOURCE: Collated from (*7, 16, 17, 19-22* Preston, C., unpublished data)

Subtle variations are observed in the ability of different resistant populations to detoxify some herbicides. For example, VLR 69 metabolizes both simazine and chlorotoluron more rapidly than does WLR 2 (*16, 17*). This is reflected in VLR 69 being slightly more resistant to PS II-inhibiting herbicides compared to WLR 2. This suggests that resistant populations can vary in the amount of cytochrome P450 monooxygenase activity that is elevated.

Studies to determine whether a single cytochrome P450 monooxygenase is responsible for metabolism-based resistance to a large number of herbicides have focussed on population VLR 69. The application of known inhibitors of cytochrome P450 monooxygenase activities to resistant plants demonstrated a high likelihood that different cytochrome P450 monooxygenases are responsible for metabolism of different herbicides in this population (*7*). For example, metabolism of both simazine and chlorotoluron are inhibited by ABT and PBO, greatly inhibited by tetcyclasis, but not inhibited by malathion. The opposite pattern of inhibition is evident for chlorsulfuron metabolism (Table V). Based on these inhibitor studies, at least four different cytochrome P450 monooxygenases are proposed to contribute to herbicide metabolism in this population. (*7*).

Table V. Effect of a 24 h Pre-Treatment of Plants with Cytochrome P450 Monooxygenase Inhibitors on Metabolism of Herbicides in *L. rigidum* Population VLR 69.

Herbicide	*Inhibitor*			
	ABT	*PBO*	*Malathion*	*Tetcyclasis*
Diclofop-methyl	++[a]	-	-	-
Tralkoxydim	-	-	-	-
Chlorsulfuron	-	-	++	-
Simazine	+	+	-	++
Chlorotoluron	+	+	-	++

[a] "-" indicates no significant impact of the inhibitor treatment, "+" indicates inhibitor reduced metabolism of the herbicide, "++" indicates inhibitor reduced metabolism of the herbicide to the rate of metabolism in susceptible plants.

SOURCE: Collated from (7).

Genetics of Multiple Resistance

Where metabolism-based resistance occurs, it frequently results in non-target site cross-resistance to a range of herbicides from many different chemical groups (Table IV). However, studies with VLR 69 (*7*), and other populations (*21, 23*), have indicated that more than one cytochrome P450 monooxygenase contributes to metabolism of herbicides in these populations. Therefore, it is possible that several cytochrome P450 monooxygenase activities may be elevated under the control of a single gene. This possibility was investigated by conducting crosses between VLR 69 and the susceptible VLR 1 to create segregating populations and examining the progeny for resistance linkages. Briefly, an F_2 population from a cross between resistant and susceptible plants was treated with high rates of simazine, chlorotoluron, tralkoxydim, or chlorsulfuron. The survivors were crossed within treatments to create selected F_3 populations. The F_3 plants were divided into four clones and each clone treated with a different herbicide. Survival of plants from each treatment were compared (Table VI).

Table VI. Survival (%) of Selected F_3 Plants, From a Cross between Resistant (VLR 69) and Susceptible (VLR 1) *L. rigidum* Populations, to Herbicides.

Selecting Herbicide (herbicide rate)	*Survival (%) to Testing Herbicide (herbicide rate)*			
	Simazine (3 kg ha^{-1})	*Chlorotoluron (2 kg ha^{-1})*	*Tralkoxydim (100 g ha^{-1})*	*Chlorsulfuron (128 g ha^{-1})*
Simazine (5.2 kg ha^{-1})	*71*[a]	68	12	16
Chlorotoluron (4 kg ha^{-1})	67	*68*	21	20
Tralkoxydim (160 g ha^{-1})	36	27	*63*	38
Chlorsulfuron (256 g ha^{-1})	32	30	47	*50*

[a] Survivorship of accessions selected with the testing herbicide is in italics.

SOURCE: (Preston, C., unpublished data)

Selection of the F_2 population with 5.2 kg ha^{-1} simazine resulted in progeny that were highly resistant to simazine with 71% surviving application of 3 kg ha^{-1} simazine. Selection of the same F_2 population with 4 kg ha^{-1}chlorotoluron produced progeny that were also highly resistant to simazine. However, progeny of selections with chlorsulfuron or tralkoxydim were not highly resistant to simazine, with only about 30% surviving treatment. In summary, selection of the segregating population with simazine results in resistance to both simazine and chlorotoluron, but not other herbicides. Similarly, selection of the population with chlorotoluron results in plants resistant to both chlorotoluron and simazine (Table VI). However, no other strong linkages between resistances were apparent, although a weak link between tralkoxydim resistance and chlorsulfuron resistance was observed. These experiments suggest that at least three genes contribute to metabolism-based resistance in VLR 69.

Glyphosate Resistance in *Lolium rigidum*

Recently glyphosate resistance was reported in two different populations of *L. rigidum* (*8, 9*). These populations had both experienced an extensive history of glyphosate use. The two populations have 8 to 10-fold resistance to glyphosate and both are also about 3-fold resistant to diclofop-methyl, but susceptible to other herbicides. One population had previously received applications of diclofop-methyl, but the other population had not. This raises the possibility that glyphosate resistance may result in diclofop-methyl cross-resistance.

The mechanism of resistance is under investigation in both populations of *L. rigidum*. In neither resistant population was substantial metabolism of glyphosate evident (*18, 24*). Absorption of glyphosate and translocation of glyphosate out of treated leaves were similar for resistant and susceptible plants. Shikimate accumulation following glyphosate application is higher in susceptible plants compared to resistant plants, suggesting that there is some barrier to penetration of gyphosate to the chloroplast in resistant plants (*18*).

More recently, several other glyphosate-resistant *L. rigidum* populations have been identified. Some of these have similar resistance to previously-described populations. However, at least one has much lower resistance to glyphosate (Figure 1), suggesting that more than one glyphosate resistance mechanism may be possible.

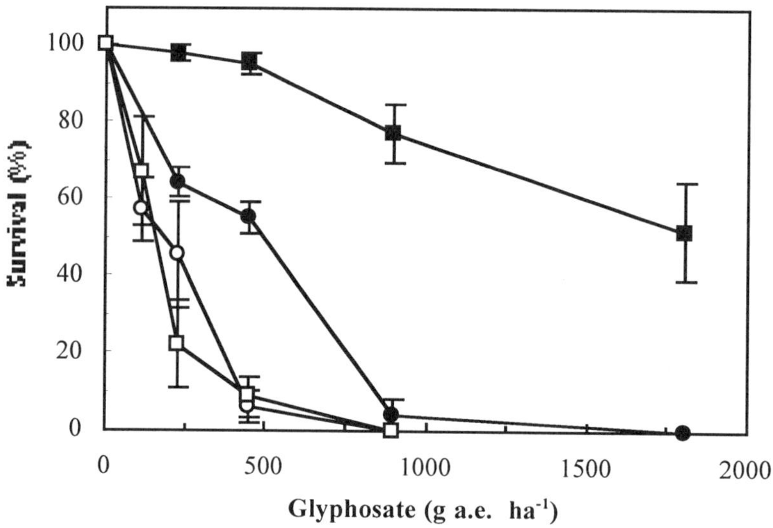

Figure 1. Dose response of two susceptible L. rigidum populations (○, □), and accessions NLR 70 (■) and 97075(●) with different levels of resistance to glyphosate (Preston, C., unpublished data).

Evolution of Multiple Resistance in *L. rigidum*

L. rigidum populations readily evolve multiple resistance to a wide range of herbicides. This occurs through the ability of *L. rigidum* populations to accumulate different resistance mechanisms. For example, population VLR 69 has both insensitive target site and enhanced metabolism-based resistance mechanisms. This population has accumulated at least five different genes encoding herbicide resistance mechanisms.

L. rigidum is a widespread weed occurring at high densities in crop fields; most of which are treated with herbicides every year. This increases the probability of selecting individuals carrying resistance genes. Indeed, the selection of populations resistant to glyphosate is testament to this. Individuals containing different resistance mechanisms can cross and produce progeny with both mechanisms. In this way, *L. rigidum* populations can accumulate a number of herbicide resistance mechanisms. These multiply-resistant populations can be resistant to most of the herbicides registered for control of *L. rigidum* in Australia and consequently pose tremendous difficulties for management.

References

1. Gill, G. S.; Holmes, J. E. *Pestic. Sci.* **1997**, *51*, 352-358.
2. Heap, J; Knight, R. J. *Aust. Inst. Agric. Sci.* **1982**, *48*, 156-157.
3. Preston, C. In *Australian Weed Mangement Systems*; Sindel, B. M., Ed.; R.G. and F.J. Richardson: Melbourne, Australia, 2000; pp 209-221.
4. Gill, G. S. *Aust. J. Exp. Agric.* **1995**, *35*, 67-72.
5. Tardif, F. J.; Holtum, J. A. M.; Powles, S. B. *Planta*, **1993**, *190*, 176-181.
6. Matthews, J. M.; Holtum, J. A. M.; Liljegren, D. R.; Furness, B.; Powles, S. B. *Plant Physiol.* **1990**, *94*, 1180-1186.
7. Preston, C.; Tardif, F. J.; Christopher, J. T.; Powles, S. B. *Pestic. Biochem. Physiol.* **1996**, *54*, 123-134.
8. Powles, S. B.; Lorraine-Colwill, D. F.; Dellow, J. J.; Preston, C. *Weed Sci.* **1998**, *46*, 604-607.
9. Pratley, J.; Urwin, N; Stanton, R.; Baines, P.; Broster, J.; Cullis, K.; Schafer, D.; Bohn, J.; Krueger, R. *Weed Sci.* **1999**, *47*, 405-411.
10. Christopher, J. T.; Powles, S. B.; Holtum, J. A. M. *Plant Physiol.* **1992**, *100*, 1909-1913.
11. Devine, M. D.; Eberlein, C. E. In *Herbicide Activity: Toxicology, Biochemistry and Molecular Biology*; Roe, R. M.; Burton, J.D.; Kuhr, R. J., Eds.; IOS Press, Inc: Amsterdam, 1997; pp 159-185.
12. Bernasconi, P.; Woodworth, A. R.; Rosen, B. A.; Subramanian, M. W.; Siehl, D. L. *J. Biol. Chem.* **1995**, *270*, 17381-17385.

13. Saari, L. L.; Cotterman, J. C.; Thill, D. C. In *Herbicide Resistance in Plants: Biology and Biochemistry*; Powles, S. B.; Holtum, J. A. M., Eds.; Lewis Publishers: Boca Raton, FL, 1994; pp 83-139.
14. Tardif, F. J.; Powles, S. B. *Physiol. Plant.* **1994**, *91*, 488-494.
15. Tardif, F. J.; Preston, C.; Holtum, J. A. M.; Powles, S. B. *Aust. J. Plant Physiol.* **1996**, *23*, 15-23.
16. Burnet, M. W. M.; Loveys, B. R.; Holtum, J. A. M.; Powles, S. B. *Planta* **1993**, *190*, 182-189.
17. Burnet, M. W. M.; Loveys, B. R.; Holtum, J. A. M.; Powles, S. B. *Pestic. Biochem. Physiol.* **1993**, *46*, 207-218.
18. Lorraine-Colwill, D. F.; Hawkes, T. R.; Williams, P. H.; Warner, S. A. J.; Sutton, P. B.; Powles, S. B.; Preston, C. *Pestic. Sci.* **1999**, *55*, 489-491.
19. Preston, C.; Powles, S. B. *Pestic. Biochem. Physiol.* **1998**, 62, 179-189.
20. Christopher, J. T.; Powles, S. B.; Liljegren, D. R.; Holtum, J. A. M. *Plant Physiol.* **1991**, *95*, 1036-1043.
21. Preston. C.; Powles, S. B. *Planta* **1997**, *201*, 202-208
22. Burnet, M. W. M. Ph.D. thesis, University of Adelaide, Adelaide, Australia, 1992.
23. Christopher, J. T.; Preston, C.; Powles, S.B. *Pestic. Biochem. Physiol.* **1994**, *49*, 172-182.
24. Feng, P. C. C.; Pratley, J. E.; Bohn, J. A. *Weed Sci.* **1999**, *47*, 412-415.

Chapter 10

Herbicide Resistance in North America: The Case for Resistance to ALS Inhibitors in the United States

D. L. Shaner[1] and I. Heap[2]

[1]Agricultural Products Research Division, American Cyanamid, P.O. Box 400, Princeton, NJ 08543–0400
[2]Weedsmart, P.O. Box 1365, Corvallis, OR 97339

Resistance to acetolactate synthase inhibitors has been documented in weed biotypes from 27 different species in the United States. These biotypes are found throughout the U.S. Most cases of resistance to ALS inhibitors are due to an alteration at the target site. Single mutations at multiple sites within the ALS gene determine the level of resistance as well as the amount of cross-resistance among different classes of inhibitors. ALS resistance can be managed by using ALS inhibitors in an integrated system that utilizes other herbicides with different mechanisms of action as well as mechanical, cultural, and biological weed control methods.

Introduction

Acetolactate synthase (ALS) (also known as acetohyrdoxyacid synthase) inhibitors were introduced into the marketplace in the 1980s (*1*). These new herbicides control a broad spectrum of broadleaf weeds and grasses at rates

© 2002 American Chemical Society

ranging from 10 to 200 g/ha. ALS inhibitors kill plants by inhibiting the first common enzyme in the branched chain amino acid pathway. Thus, these herbicides do not affect mammals, fish, birds or insects, at least in part, because this target does not exist in these organisms (*1*).

There are over 30 commercial ALS inhibitors used in multiple crops throughout the world (*2*). These herbicides are derived from four major chemical classes including imidazolinones, sulfonylureas, triazolopyrimidine sulfonamides, and pyrimidinylsalicylates (Figure 1) (1). The ALS inhibitors accounted for approximately 17.5% of the total world herbicide market in 1997 (*3*).

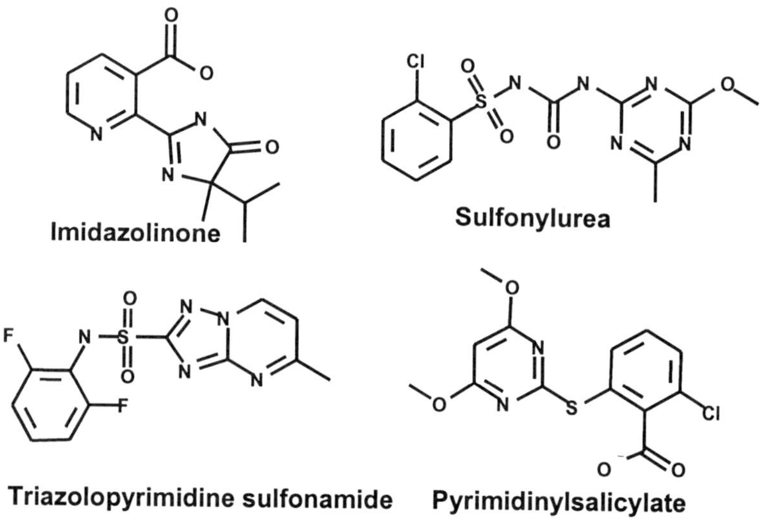

Figure 1: Examples of different classes of ALS Inhibitors

Development of Resistance to ALS Inhibitors in the U.S.

In 1982 chlorsulfuron was registered in the U.S. for use in cereals (*4*). The wheat growers in the northwestern U.S quickly adopted this highly effective, and long residual herbicide. In 1987 a farmer in Idaho found that he was no longer controlling patches of prickly lettuce (*Lactuca serriola*) in fields that had received annual applications of chlorsulfuron for 5 years (*4*) In 1988 and 1989 chlorsulfuron resistant populations of kochia (*Kochia scoparia*) and Russian thistle (*Salsola iberica*) were found throughout the cereal growing area of the U.S. where chlorsulfuron was the only herbicide that had been used in these fields for 3 to 5 years (*4*).

In 1992 populations of smallflower umbrella sedge (*Cyperus difformis*) and California arrowhead (*Sagittaria montevidensis*) with resistance to bensulfuron, a widely used sulfonylurea in rice, were discovered in California (*4*). In Mississippi a population of cocklebur (*Xanthium strumarium*) that had received two applications of imazaquin per year for a 4-year period was no longer controlled by this herbicide (*4*). At present there are over 27 species in which resistant biotypes have been selected to a number of different ALS inhibitors in the U.S. (*5*).

Resistance to ALS inhibitors can be found throughout the U.S. in wheat, corn, soybeans, rice, and along railways and roadsides (Figure 2) (*5*). The distribution of ALS resistance is highly correlated with the use pattern of these herbicides. For example, ALS inhibitors were registered for use in soybeans in the late 1980s. Due to their high efficacy and cost effectiveness, there was a steady increase in the soybean acreage treated until over 80% of the soybeans were treated in 1996 (Figure 3). The increase in the reported cases of ALS inhibitor resistant weeds also increased over this same time period except with a 6-year lag. In most cases, resistance to ALS inhibitors takes 3 to 6 years of continuous use before resistance appears, which could explain the time lag. The majority of the cases of resistance to ALS inhibitors have been selected in corn/soybean growing areas (*5*). Interestingly, the extent of soybeans treated with ALS inhibitors dramatically decreased with the introduction of glyphosate-resistant soybeans (Figure 3). The judicious use of glyphosate in combination with the ALS inhibitors may provide a powerful tool for managing resistance to both these herbicide classes. With this new tool the rate of increase of ALS resistant weed biotypes in the U.S. may decrease.

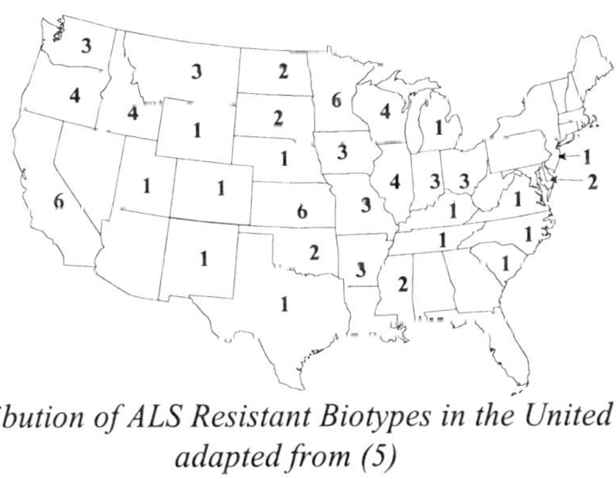

Figure 2: Distribution of ALS Resistant Biotypes in the United States. Data adapted from (5)

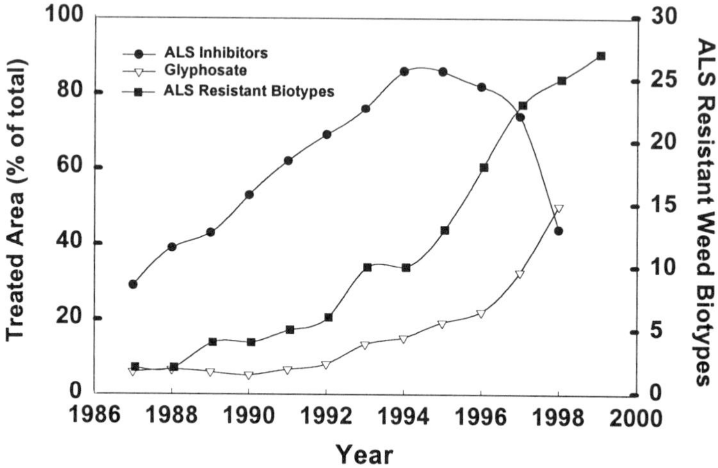

Figure 3: Soybean area treated with ALS inhibitors, glyphosate and the appearance of ALS Inhibitor Resistant Weed Populations in the U.S. Data adapted from (5,6).

Mechanisms of Resistance

In most of the populations where the mechanism of resistance has been determined, resistance to ALS inhibitors was due to the selection of an altered ALS enzyme that was no longer sensitive to the herbicides. Although there are at least 10 different sites within 3 conserved regions of the ALS gene where mutations can result in a resistant enzyme (*7*), most mutations occur in one of four sites (Table 1). The most common site, particularly for biotypes selected by a sulfonylurea was at a proline (Pro) site where a mutation that substitutes any other amino acid for proline resulted in resistance (*4*). Guttieri et al. (*8*) found that in 8 populations of ALS resistant kochia studied, 6 of the populations contained a mutation at this proline site. Cross-resistance studies have shown that mutations at the Pro site resulted primarily in resistance to sulfonylureas and triazolopyrimidines (*9*). At another site where a mutation changes a tryptophan (Trp) to leucine (Leu), the plant becomes broadly cross-resistant to all ALS inhibitors (*10*).

Table 1: Resistance Pattern of Different Mutations in ALS to Inhibitors

Amino Acid	*Position*	*Mutation*	*Resistance*	*Reference*
Pro	197	Ser, His, Leu, Gln, Ala, Thr	Sulfonylurea Triazolopyrimidine	*9*
Ser	653	Asp	Imidazolinone	*11*
Ala	56	Thr	Imidazolinone Pyrimidylsalicylate	*10*
Trp	552	Leu	Imidazolinone Sulfonylurea Triazolopyrimidine Pyrimidylsalicylate	*10*

There appear to be few differences between ALS inhibitor resistant biotypes and wild type biotypes. One difference that has been noted is on germination. In some cases, ALS inhibitor resistant biotypes produce seeds that germinate more rapidly and at lower temperatures compared to wild type seed (*12, 13*). Analysis of the pool sizes of the branched chain amino acids in two kochia and prickly lettuce biotypes showed that the resistant biotypes had higher levels of valine (Val), leucine (Leu), and isoleucine (Ile) than wild type seed (*13, 14*). Work by Eberlein et al (*14)* also found that the feedback sensitivity of ALS by Val and Leu was less in the resistant lines of prickly lettuce compared to the sensitive lines. They proposed that the differences in the branched chain amino acid pools was due to this lower feedback sensitivity. However, the relationship between these differences in amino acid pool sizes and germination remains unknown.

Growth and competition studies on resistant (R) and susceptible (S) prickly lettuce biotypes under greenhouse conditions showed that the S biotypes produced 31% more biomass than the R biotypes, but the relative competitive ability of the two biotypes was similar (*12*). R and S biotypes of kochia were also equally competitive. R and S kochia and prickly lettuce biotypes produced similar amounts of seed, and the seed longevity of R and S prickly lettuce biotypes was similar under field conditions (*12)*.

The catalytic efficiency of ALS appears to be unaffected by the mutations that result in ALS inhibitor resistance. The Km for pyruvate in ALS from resistant kochia, wild lettuce, Russian thistle, common chickweed (*Stellaria media*), and perennial ryegrass (*Lolium perenne*) ranged from 1.7 to 4.8 mM, which is close to Km values reported for susceptible plant ALS (*4, 14*).

There are a few cases where ALS inhibitor resistance appeared to be due to an alteration of the rate of metabolism of the herbicides. Mallory Smith et al (*15*) described a biotype of downy brome (*Bromus tectorum)* that was selected through annual use of primisulfuron and that was cross-resistant to sulfosulfuron. The mechanism of resistance in this population appeared to be due to altered metabolism and not to an altered target site. In this case, the R biotype may be able to more rapidly detoxify the herbicides than S biotypes. This was the first documented case where resistance due to altered metabolism was selected by the use of an ALS inhibitor.

Management of ALS Inhibitor Resistance

Although ALS inhibitor resistant biotypes are widely distributed throughout the U.S., the actual acreage infested with these R biotypes is still relatively small compared to the total acreage treated with these herbicides. In most cases R biotypes have been selected after 3 to 8 years of continuous use of the ALS inhibitors. Resistance has not been selected where the ALS inhibitors are part of an integrated program that included other herbicides with different mechanisms of action that control a similar weed spectrum as the ALS inhibitor, combined with mechanical, cultural, and biological control methods. (*16*).

Thus, farmers should avoid using an ALS inhibiting herbicide as the sole means of weed control in any field over a long period of time. These herbicides should be part of an integrated weed management program that includes other weed control methods. If ALS inhibitors are used in combination with other herbicides, it is important that the two herbicides control the same spectrum of weeds. It does no good, from a resistance management perspective, to mix herbicides that control different spectrums of weeds. In addition, one should tailor the herbicide program to the weed spectrum in a field and use the minimum amount of herbicide necessary to control the weeds to avoid using herbicides in excess. Finally one should use certified crop seed and clean equipment when moving from one field to another to prevent spreading resistant weed seed or plant material.

References

1. Shaner, D. L.; Singh, B. K. In *Herbicide Activity, Toxicology, Biochemistry, and Molecular Biology;* Roe, R. M.; Burton, J. D.; Kuhr, R. J. Eds.; IOS Press: Amsterdam, 1997; pp. 69-110.

2. Saari, L. L. In *Pesticide Chemistry and Bioscience: The Food-Environment Challenge;* Brooks, G.T.; Robers, T. R., Eds.; Royal Society of Chemistry, Cambridge, 1999 pp. 207-220.
3. Matthew, P.; Mathiesen, F.; Galloway, F.; McDougall, *J. Agrochemical Service* Wood Mackenzie Consultants, Ltd. London, **1998** pp.12-47.
4. Saari, L. L.; Cotterman, J. C.; Thill, D. C. In *Herbicide Resistance in Plants: Biology and Biochemistry*;. Powles, S. B.; and Holtum, J. A. M., Ed.s; Lewis Publishing: Boca Raton, FL, 1994; pp. 83-139.
5. Heap, I. International Survey of Herbicide Resistant Weeds. http://www.weedscience.com. (Accessed 1 Sept 1999)
6. Shaner, D. L. *Pest. Sci.* **2000,** (in press).
7. Falco, S. C.; Dumas, K. S. *Genetics.* **1985**, *109*, 21-35.
8. Guttieri, M .J.; Eberlein, C. V.; Thill, D. C. *Weed Sci.* **1995**, *43*, 175-182.
9. Saari, L. L; Mauvis, C. J. In *Herbicide Resistant Crops;*.Duke, S.O., Ed.; Lewis Publishers: Boca Raton, FL, 1996; pp. 127-142.
10. Bernosconi, P.; Woodworth, A. R.; Rosen, B. A.; Subramanian, M. V. *J. Biol. Chem.* **1995**, *270*, 17381-17385.
11. Sathasivan, K.; Haugh, G. W.; Murai, N. *Plant. Physiol.* **1991,** *97,* 1044-1050.
12. Mallory-Smith, C. A.; Thill, D. C.; Alcocer-Ruthling, M.; Thompson, C. *Proc. of the First International Weed Control Congress*, **1992,** *2*, 301-303.
13. Dyer, W. E.; Chee , P. W.; Fay, P. K. *Weed Sci.* **1993** *41*, 18-22.
14. Eberlein, C. V.; Guttieri, M. J.; Berger, P. H.; Fellman, J. K.; Mallory-Smith, C. A.; Thill, D. C.; Baerg, R. J.; Belknap, W. R. *Weed Sci.,* **1999,** *47,* 383-392.
15. Mallory-Smith, C.; Hendrickson, P.; Mueller-Warrant, G. *Weed Sci.* **1999,** *47,* 256-257.
16. Shaner, D. L.; Feist, D. A.; Retzinger, E. J. *Pest. Sci*: **1997**, *51*, 367-370.

Chapter 11

Sulfonylurea-Resistant Weeds in Paddy Rice Fields of Japan

Kazuyuki Itoh[1] and Akira Uchino[2]

[1]National Institute of Agro-Environmental Sciences, Kannondai, Tsukuba 305–8604, Japan

[2]National Agricultural Research Center for Tohoku Region, Omagari, Akita 014–0102, Japan

Resistance to sulfonylureas developed in nine annual weeds and one perennial weed found in rice cultivation areas of northern and central Japan. Some resistant populations demonstrate high sulfonylurea resistance. Fifty percent lethal doses (LD50) of some resistant biotypes were approximately 100 to 1000 times higher than those of susceptible biotypes. A rapid diagnostic method was applied to the resistant weeds, and herbicide treatment with thifensulfuron-methyl {methyl 3-[[[[(4-methoxy-6-methyl-1,3,5-triazin-2-yl)=amino]carbonyl]amino]sulfonyl]-2-thiophenecarboxylate } under adequate illumination yielded clear results. A crossing test suggested that dominant nucleic gene(s) control the inheritance behavior of the sulfonylurea resistance. Genetic variations and gene mutation were investigated in the resistant weeds. A control method was confirmed for sulfonylurea resistance weeds in paddy fields in Japan.

In the early 1980s, K. Itoh and Dr. Y. Watanabe began studying paraquat-resistant (see chemical names of herbicides in Table I) *Erigeron*

 © 2002 American Chemical Society

Table I. Common and chemical names of herbicides cited in this section

Common Name	*Chemical Name*
benfuresate	2,3-dihydro-3,3-dimethyl-5-benzofuranylethanesulfonate
bensulfuron-methyl	methyl 2-[[[[[(4,6-dimethoxy-2-pyrimidinyl)amino]=carbonyl]amino]sulfonyl]methyl]benzoate
benzofenap	2-[[4-(2,4-dichloro-3-methylbenzoyl)-1,3-dimethyl-=1*H*-pyrazol-5-yl]oxy]-1-(4-methylphenyl)ethanone
bromobutide	2-bromo-3,3-dimethyl-*N*-(1-methyl-1-phenylethyl)=butanamide
chlorsulfuron	2-chloro-*N*-[[(4-methoxy-6-methyl-1,3,5-triazin-2-yl)=amino]carbonyl]benzenesulfonamide
dimepiperate	*S*-(1-methyl-1-phenylethyl) 1-piperidinecarbothioate
esprocarb	*S*-(phenylmethyl) (1,2-dimethylpropyl)=ethylcarbamothioate
ethoxysulfuron	1-(4,6-dimethoxypyrimidin-2-yl)-3-(2-ethoxyphenoxysulfonyl)urea
imazosulfuron	2-chloro-*N*[[(4,6-dimethoxy-2-pyrimididyl)amino]=carbonyl]imidazo[1,2-α]pyridine-3-sulfonamide
MCPB	4-(4-chloro-2-methylphenoxy)butanoic acid
mefenacet	2-(2-benzothiazolyloxy)-*N*-methyl-N-phenylacetamide
metsulfuron-methyl	methyl 2-[[[[(4-methoxy-6-methyl-1,3,5-triazin-2-yl)=amino]carbonyl]amino]sulfonyl]benzoate
paraquat	1,1'-dimethyl-4,4'-bipyridinium
pentoxazone	3-[4-chloro-5-(cyclopentyloxy)-2-fluorophenyl]-=5-(1-methylethylidene)-2,4-oxazolidinedione
pretilachlor	2-chloro-*N*-(2,6-diethylphenyl)-N-(2-propoxyethyl)=acetamide
pyrazolate	(2,4-dichlorophenyl)[1,3-dimethyl-5-[[(4-methyl=phenyl)sulfonyl]oxy]-1*H*-pyrazol-4-yl]methanone
pyrazosulfuron-ethyl	ethyl 5-[[[[(4,6-dimethoxy-2-pyrimidinyl)amino]=carbonyl]amino]sulfonyl]-1-methyl-1*H*-pyrazole-4-carboxylate
pyributicarb	*O*-[3-(1,1-dimethylethyl)phenyl] (6-methoxy-=2-pyridinyl)methylcarbamothioate
simetryn	*N*,*N*'-diethyl-6-(methylthio)-1,3,5-triazine-2,4-diamine
thifensulfuron-methyl	methyl 3-[[[[(4-methoxy-6-methyl-1,3,5-triazin-2-yl)=amino]carbonyl]amino]sulfonyl]-2-thiophenecarboxylate
thiobencarb	*S*-[(4-chlorophenyl)methyl] diethylcarbamothioate

philadelphicus L., a perennial weed in the Compositae family in mulberry fields in Saitama Prefecture in Japan. Subsequently, some paraquat-resistant weeds in Compositae and in other families were found in fields of fruit trees, mulberries, vegetables, and tea, and in paddy levees in Japan and Malaysia (*1, 2*).

Sulfonylurea (SU) herbicides were introduced in the early 1980s and have since been widely used in major cereal-growing areas to control or suppress broad leaf weeds and sedges. Several weed species have developed resistance to SU herbicides as a result of repeated use of the same SU-based herbicides. The first two cases of SU resistance were confirmed in *Kochia scoparia* L. (*3*) and *Lactuca serriola* L. (*4*) after several years of selection by chlorsulfuron. The site of SU action is thought to be acetolactate synthase (ALS). The number of SU resistant weeds, in addition to weeds resistant to other ALS-inhibiting herbicides, has reached more than 60 (*5*). Most SU-resistant weeds were collected from a field where winter wheat was grown and where chlorsulfuron or chlorsulfuron plus metsulfuron-methyl had been applied for three to five years. SU-resistant paddy weeds have also been collected since 1992 in the *Cyperus difformis* L. and *Sagittaria montevidensis* L. subsp. *Calycina* from rice fields where bensulfuron-methyl (BSM) was applied in the USA and Australia (*6*).

Before SU herbicides were introduced, farmers in rice cultivation in Japan generally applied herbicides into paddy fields twice in a cropping season by using herbicides with different modes of action. Farmers began to apply herbicides only once a season after the introduction of SU, using herbicides that included SU mixed with mefenacet or esprocarb. SU-resistant weeds have appeared under these conditions in paddy fields in Japan. We describe these SU-resistant weeds in this report.

Finding of SU-Resistant Weeds in Japan

BSM and pyrazosulfuron-ethyl (PSE) were used in paddy fields for more than ten years in Japan as ingredients of "one-shot-treatment herbicides", e.g. Granules of BSM/mefenacet, BSM/esprocarb, BSM/dimepiperate, and PSE/mefenacet (Zark, Fujigrass, Push, and Act are their respective commercial names in Japan).

In 1995, *Monochoria korsakowii* Regel et Maack (Japanese name: Mizuaoi), a Pontederiaceae paddy weed, was first discovered to be resistant to SU herbicides in Hokkaido Prefecture, Japan (*7, 8*). An SU-resistant biotype of this species was discovered in Korea in 1998, in reclaimed rice fields in the west coastal area, where SU herbicides had been used for nine years (*9*). In spring 1996, annual paddy weeds in the family of Scrophulariaceae, *Lindernia micrantha* D. Don. (Japanese name: Aze-togarashi) and *L. dubia* Pennell var. *dubia* (Japanese name: Taketo-azena), were observed to be SU resistant at

Kawanishi and Yuza Town, respectively, in Yamagata Prefecture (*10, 11, 12, 13*). We also found SU-resistant plants in *L. dubia* var. *major* Pennell (Japanese name: Amerika-azena) in some paddy fields in Akita and Miyagi Prefectures (*11, 12*). Both varieties of *L. dubia* are native to North America. *L. procumbens* (Krock.) Philcox. (Japanese name: Azena) and *Limnophila sessiliflora* Blume (Japanese name: Kikumo) were also unable to be controlled by either SU herbicides or barnyard grass-killers in the Tohoku and Hokuriku districts (*11, 12, 14, 15*).

Scirpus juncoides Roxb. var. *ohwianus* T. Koyama (Japanese name: inuhotarui), a serious perennial weed that severely infested paddy fields in Hokkaido and Miyagi Prefectures, was recently found to be resistant to SU herbicides (*16, 17*). Kohara et al (*16*) characterized the fields infested with resistant weeds as follows: 1) herbicides had been properly used, 2) the other weeds had been completely controlled by herbicides, 3) severe infestation of *S. juncoides* had occurred for more than two years, 4) *S. juncoides* had been completely controlled before the severe infestation occurred, and 5) SU herbicides had been used for more than five years.

Rotala indica Koehne (Japanese name: Kikashigusa) in Akita Prefecture (*18*), *Elatine triandra* Schk. in Saitama Prefecture (*19*), and *Monochoria vaginalis* (Burm. f.) Kunth. (Japanese name: Konagi) in Akita and Ibaraki Prefectures (*20*), which are common annual paddy weeds, were found to be resistant to SU herbicides. These ten species (Table II) with SU herbicide resistance are spreading and currently cause problems in Japan.

Field Surveys in Severely Infested Areas

A field survey using a questionnaire for research with farmers was conducted in the rice growing season in the middle of June 1996, at Yuza Town of Yamagata Prefecture, where SU-resistant biotypes of *Lindernia* weeds (mostly *L. dubia* var. *dubia*) had vigorously infested fields. The weeds were observed in 229 (34.1%) in a total of 671 patches (189 farmers, 250 ha); the survey was conducted two to three weeks after herbicide treatments. The infested paddy fields were distributed separately, not concentrically. The distribution pattern was dependent on the weed control method that had been implemented in the fields by individual farmers, and the infestations were related to the consecutive use of one-shot-treatment herbicides, including SU; for example, BSM or PSE. Consecutive applications of the herbicides may have intensified the infestation of the SU-resistant biotypes (*10*).

Based on the agricultural background of rice cultivation in the surveyed areas, the causes of infestation of *Lindernia* spp. in rice fields can be considered to be 1) the consecutive use of one-shot-treatment herbicides, 2) an expansion of

Table II. Weed species resistant to sulfonylurea herbicides in the paddy fields of Japan

Scientific Name	*Japanese Name*	*Finding Year*	*Finding Prefecture*	*References*
Monochoria korsakowii	Mizuaoi	1995	Hokkaido	7,8,28
Lindernia micrantha	Azetogarashi	1996	Yamagata	13,24,26,29, 30
		1997	Saitama	21
		1998	Kyoto	22,29
		1998	Akita	29
L. dubia var. *dubia*	Taketo-azena	1995	Yamagata	10,26,30
		1995	Miyagi	12
		1996	Saitama	19,21
L. dubia var. *major*	Amerika-azena	1995	Akita	12,33
		1996	Aomori	15
		1996	Miyagi	12,26,30
		1996	Saitama	19,21
L. procumbens	Azena	1995	Yamagata	12
Limnophila sessiliflora	Kikumo	1996	Akita	23,33
Rotala indica	Kikashigusa	1997	Akita	18
Elatine triandura	Mizohakobe	1996	Saitama	19,21
Scirpus juncoides	Inuhotarui	1997	Hokkaido	16,34,35
		1998	Miyagi	17,31
M. vaginalis	Konagi	1999	Akita	20
		1999	Ibaraki	20

water-seeded rice cultivation, 3) the presence of a partially unplanted fallow area in paddy fields imposed by rice production regulations, 4) seed movement related to agricultural machines, and 5) soil movement that occurs with paddy consolidation. The same conditions as present in Yuza Town were observed in Nangai Village of Akita Prefecture, Kawanishi Town of Yamagata Prefecture (*12, 13*), Kazo City of Saitama Prefecture (*21*), and Omiya Town of Kyoto Prefecture (*22*), which are in separate districts in the central and northern areas of Japan.

Response to SU Herbicides

The seeds of resistant *L. micrantha* were collected from the plants that had survived SU-based herbicide treatments in a paddy field of Kawanishi Town of Yamagata Prefecture, and the seeds of a susceptible biotype were collected in an untreated area in Omagari City of Akita Prefecture in 1996. The plants of *L. micrantha* in Kawanishi Town exhibited a high level of resistance to BSM, and showed cross resistance to PSE, imazosulfuron (IM), and ethoxysulfuron (ES). The 50% lethal dose (LD50) values for resistant *L. micrantha* varied, at 105 g a.i./ha for PSE, 719 g a.i./ha for BSM, and 804 g a.i./ha for IM, which were 141, 282, and 81 times higher than those of the susceptible biotypes (*13*). The same experiment was conducted with *L. sessiliflora* that had survived SU-based herbicide treatments in a paddy field in Sennan Village of Akita Prefecture in 1997. The resistant *L. sessiliflora* also exhibited a high level of resistance to BSM, and showed cross-resistance to PSE, IM, and ES. The LD50 values for resistant *L. sessiliflora* also varied, at 110 g a.i./ha for PSE, 1,316 g for BSM, and 1,245 g for IM, which are 334, 655, and 896 times higher than those of the susceptible biotypes (*23*). It was concluded that the resistant biotypes of these Scrophulariaceae weeds are 100 to 1000 times more resistant to four kinds of SU herbicides than are the susceptible ones (*13*).

The variations of LD50 values for the SU-resistant biotypes were also observed in *S. juncoides* sampled from Iwamizawa City, Nakafurano Town, and Kuriyama Town in Hokkaido Prefecture, where the LD50 values were 40 to 80 (PSE), 75 to 90 (IM), and 55 to140 (BSM) times higher than those of susceptible biotypes sampled from the experimental paddy field in Hokkaido Central Agricultural Experiment Station (*16*).

Seasonal Change of Emergence

A seasonal change of emergence was investigated in an SU-resistant biotype of *L. micrantha* in 50× 50× 30cm concrete pots. The experiment was conducted in the rice-growing season from early May to mid-September of 1998 and 1999. Fifty thousand seeds of SU-resistant *L. micrantha* from Kawanishi Town were sown in pots under well-drained paddy conditions (well-drained from May 17 to July 2 and submerged from July 10 to Sept. 10) and poorly-drained paddy conditions (submerged throughout the year) in autumn 1997 in Tsukuba City of Ibaraki Prefecture. The rice seedlings were transplanted in the middle of May after puddling of the soil. BSM/mefenacet granules were applied at 30kg/ha one week after transplanting. All emerged weeds were pulled and counted once a week in a 25× 25cm quadrant set in the center of the concrete pot.

The seasonal changes of emergence from buried seeds of SU-resistant *L. micrantha* are shown in Figures 1 and 2. Seedlings emerged very well under

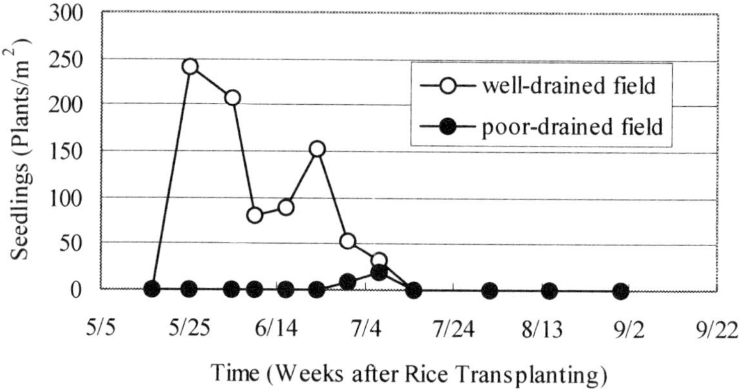

Figure 1. Seasonal change of Lindernia micranha resistant to SU herbicides in 1998 (24)

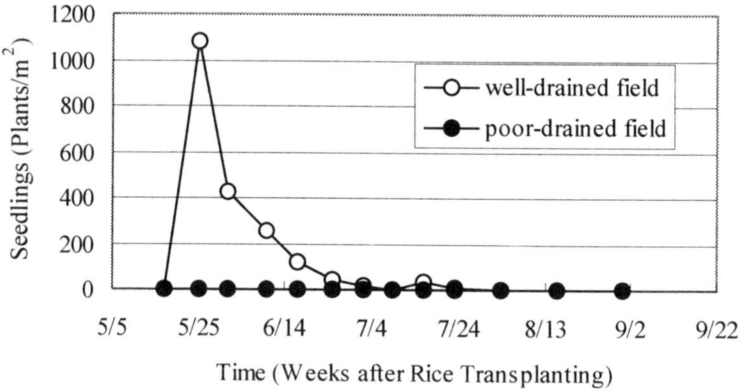

Figure 2. Seasonal change of Lindernia micranha resistant to SU herbicides in 1999 (24)

well-drained paddy conditions, but they did not emerge well under poorly-drained paddy conditions in 1998 or 1999. A few seedlings emerged after rice harvesting time under well-drained conditions. The seeds of *L. micrantha* germinated only from the paddy soil surface. This indicates that *L. micrantha* seeds rarely emerge out of the soil under reduced conditions, 2 to 3 cm below a submerged soil surface. The submerged conditions did not disturb the plant growth after the seedlings had emerged. The paddy fields in Japan were restructured into large-scale paddy fields for land conservation after World War II. Tiles were buried in the fields for water drainage during the reconstruction. Thus, there has been a transition from poorly-drained paddy fields to well-drained paddy fields in Japan. This transition may make *L. micrantha* grow easily in paddy fields (*24*)

Rapid Diagnosis of SU-Resistant Weeds

We must know the biotypes of the weeds grown in each paddy to utilize the appropriate control method for resistant weeds. There is no significant difference between the appearances of resistant and susceptible biotypes. We therefore applied the rapid diagnosis method reported by Gerwick et al. (*25*) to *Lindernia* weeds in Japan. The main process of this diagnosis method is the assay of ALS activity by using the colorimetric method. Light illumination was needed in our experiment for the assay process to diagnose *Lindernia* weeds (*26*). This light requirement may be due to the greater need for substrate for ALS in *Lindernia* weeds, since light was not needed in a subsequent experiment when adequate sucrose was given to the diagnosed tissue as a substitute for photosynthate.

Resistant weeds can be identified visually in the diagnosis by the red color of the final solution. This red color is the result of the colorimetry of acetoin accumulation in the diagnosed tissue. Acetoin accumulation is dependent on ALS activity under the conditions used in the diagnosis. ALS of a susceptible biotype is inhibited by an SU herbicide and the red color cannot be seen in tissue treated with an SU herbicide. If the ALS activity is not inhibited in the diagnosed tissue, even if it has been treated with an SU herbicide, the red color is observed in the resulting solution and the tissue can be diagnosed as SU resistant.

We used thifensulfuron-methyl as a sulfonylurea herbicide in the diagnosis. The dominant sulfonylurea herbicide used in Japanese rice production is BSM, but it is only sold as an ingredient in combination herbicides and is not sold as a wettable powder formula. Moreover, BSM has low solubility in water. Thifensulfuron methyl is one of the few sulfonylureas commercially available as a wettable powder that contains no other herbicidal ingredient, and it has a high solubility in water. A clear result was obtained in the diagnosis after using thifensulfuron-methyl at 7.5 or 75 ng a.i./ml; the resistant biotypes produced a red color while the susceptible biotypes produced a brown or yellow color in the final solution (*26*).

Diffusion and Inheritance of SU-Resistant Genes

To clarify the dispersal process of SU-resistant genes from paddy weeds of *Lindernia*, the species and activities of insects visiting the flowers of the weeds were surveyed in fallow rice fields in Kawanishi Town and Nangai Village. *Lasiogloum scitulum* (Halictidae, Hymenoptera) and *Sphaerophoria macrogaster* (Syrphidae, Diptera) were the major visitors from a number of individuals among five bee types and two hover flies collected on the flowers. The most important pollinator of *Lindernia* flowers was considered to be *L. scitulum*, based on the observation of flower-visiting behavior. A comparison between two *Lindernia* flowers showed that *L. micrantha* is a cross-pollination type and *L. dubia* var. *major* is a self-pollination type (*27*).

An SU-resistant biotype of *L. micrantha* from Kawanishi Town was crossed with a susceptible biotype from Omagari City to investigate the inheriting of an SU-resistant gene. The results are shown in Table III. All seedlings from the F_1 progenies resulting from S×S died after being treated with PSE, although some progenies from crosses S×R, R×S, and R×R survived. This suggests that dominant nucleic genes control the inheritance of the SU-resistant *L. micrantha*. The number of survivors in each line varied significantly because 1) the seedlings were very small, 2) the soil surface in the pots was exfoliated, 3) algae developed, and 4) *Daphnia* spp. and small aquatic insects were present and stirred the soil surface. The remaining S×R progenies were grown after PSE treatment, and the pollen grains were crossed with susceptible ovules. The results provided good evidence that a single dominant gene controls the SU resistance in *L. micrantha* (Table IV). Some results of S×F_1(R) (line 7) and F_1(R) ×F_1(R) (lines 9 and11) in Table IV did not satisfy the x^2 test. This may be due to the same factors as described above (*24*).

A crossing test of SU-resistant and susceptible biotypes of *M. korsakowii* was also conducted. All seedlings from the F_1 progenies resulting from S×S died after being treated with BSM, although all progenies from crosses S×R, R×S, and R×R survived. This suggests that dominant nucleic genes control inheritance of the SU-resistant gene in *M. korsakowii.* Both parents were homozygous in the experiment. The gene flow was also investigated in an experimental population of SU-resistant *M. korsakowii* in Omagari City. A bee, *Apis*, was one of the main pollinators of *M. korsakowii.* Many insects came to the flowers, and our cross ratio showed 10-65% (*28*).

Table III. Segregation ratio in F1 progenied of 4 kind of crosses between sulfonylurea resistant (R) and susceptible (S) biotypes of *Lindernia micrantha.*

Parental biotype ♀×♂	*Total No. of Seedlings*[a]	*No. of Survivals*[b]
S×S	303	0
S×S	184	0
S×S	107	0
S×R	23	21
S×R	67	27
S×R	66	14
S×R	107	27
R×S	51	49
R×S	36	35
R×S	127	3
R×R	58	33
R×R	53	11
R×R	56	4

[a] Feb.6, 1998 Seeded, Feb. 19, Observed

[b] Feb. 20, Treated PSE, April 7, Observed

Genetic Variation and Gene Mutation

Genetic variations were investigated in resistant and susceptible biotypes of *L. micrantha* by using molecular markers to clarify the process of evolution to SU resistance. The results indicated that the populations of resistant biotypes contained fewer variations than those of susceptible biotypes, and in some cases the populations of resistant biotypes were nested separately in the populations of susceptible biotypes collected in separate regions. Therefore, some populations of resistant biotypes occurred spontaneously in different regions in Japan (*29*).

Uchino and Watanabe (*30*) reported mutations of the ALS gene of *Lindernia* spp. A comparison of the deduced amino acid sequences of domain A, which is usually a highly conserved region of ALS, showed that amino acid substitution occurred in all resistant biotypes in the proline residue of domain A,

Table IV. Segregation ratio in back cross and F1 progenies of the cross between each survived plants of S × R

Parental biotype ♀×♂	*Total No. of Seedlings*[a]	*No. of Survival*[b]	*No. of Dead Seedlings*[b]	*Expected ratio (R:S)*[c]	X^2	*P*
S ×F_1 (R)	246	107	139	1:1	4.163	>0.01
S ×F_1 (R)	165	85	80	1:1	0.152	>0.05
S ×F_1 (R)	97	43	54	1:1	1.247	>0.05
S ×F_1 (R)	187	104	83	1:1	2.356	>0.05
S ×F_1 (R)	168	87	81	1:1	0.214	>0.05
S ×F_1 (R)	317	138	179	1:1	5.303	>0.01
S ×F_1 (R)	351	116	235	1:1	40.345	<0.001
F_1 (R) ×F_1 (R)	86	70	16	3:1	2.302	>0.05
F_1 (R) ×F_1 (R)	187	104	83	3:1	37.480	<0.001
F_1 (R) ×F_1 (R)	198	150	48	3:1	0.061	>0.05
F_1 (R) ×F_1 (R)	262	159	103	3:1	28.625	<0.001

[a] Oct. 30, 1998 Seeded, Nov. 14, Observed

[b] Nov. 15, BSM Treated, Dec. 25, Observed

[c] Single dominant gene

while the proline was conserved in all susceptible biotypes. Shibuya et al. (*31*) also reported a mutation of ALS in domain A in resistant biotypes of *S. juncoides*.

How to Control the Resistant Species

A sequential application of an herbicide for soil application (400 g a.i./ha of pretilachlor; commercially named Solnet in Japan) and an herbicide for foliar and soil application (240 g a.i./ha of MCPB, 450g a.i./ha of simetryn, and 3.0 kg a.i./ha of thiobencar; commercially named Kumilead-SM in Japan) is effective for SU-resistant annual paddy weeds (*2, 13, 32*). Pentoxazone is also effective for SU-resistant annual broad leaves, such as *Lindernia* spp (*33*).

We must utilize some one-shot-treatment herbicides, such as bromobutide and SU (*34*), or pyrazolate, pretilachlor, and SU (*35*), to control the resistant biotypes of *S. juncoides*. Pretilachlor/ benfuresate/ pyrazolate/ dimepiperate (a four-compound mixed herbicide) or pyributicarb/ bromobutide/ benzofenap (a three-compound mixed herbicide), herbicides composed of ingredients with different modes of action, are also effective on the resistant *S. juncoides* (*16*).

References

1. Itoh, K.; Matsunaka, S. In *Biological Approaches and Evolutionary Trends in Plants;* Kawano, S., Ed.; Academic Press, 1990; pp 33-49.
2. Itoh, K. In *Proc. of Invasion of Herbicide Resistant Weeds in Agro-ecosystems in Japan;* Itoh, K., Ed.; National Institute of Agro-Environmental Sciences: Tsukuba, Japan, 1998; pp 82. (in Japanese)
3. Primiani, M. M.; Cotterman, J. C.; Saari, L. L. *Weed Technol.* **1990,** *4,* 169 172.
4. Mallory-Smith, C. A.; Thill, D. C.; Dial, M. J. *Weed Technol.* **1990,** *4,* 163-168.
5. Heap, I. *International Survey of Herbicide Resistant Weeds,* URL http://www.weedscience.com/
6. Saari, L. L.; Cotterman, J. C.; Thill, D.C. In *Herbicide Resistance in Plants: Biology and Biochemistry;* B. Powles; Holtum, J. A. M., Eds.; Lewis Publishers: Boca Raton, FL, 1994; pp 83-139.
7. Kohara, H.; Yamashita, H.; Yamazaki, N. *Weed Res. Japan* **1996,** *41(suppl.),* 236-237. (in Japanese)
8. Wang, G. X.; Kohara, H.; Itoh, K. In *1997 Brighton Conference-Weeds-Proceedings;* The British Crop Protection Council, Ed,; Brighton, UK, 1997; 311-318.
9. Park, T. S.; Kim, C. S.; Park, J. P.; Oh, Y. K.; Kim, K. U. In *Proc. 17th Asian-Pacific Weed Science Society Conf.;* Bangkok, 1999; pp 251-254.
10. Itoh, K.; Uchino, A.; Wang, G. X.; Yamakawa, S. *J. Weed Sci. Tech.* **1997,** *42(suppl. 1),* 22-23. (in Japanese)
11. Itoh, K.; Wang, G. X. In *Proc. 16th Asian-Pacific Weed Science Society Conf.* Kuala Lumpur, Malaysia, 1997; pp 219-221.
12. Itoh, K.; Wang, G. X.; Uchino, A. *J. Weed Sci. Tech.* **1997,** *42(suppl. 1),* 12-13. (in Japanese)
13. Itoh, K.; Wang, G. X.; Oba, S. *Weed Res.* **1999,** *39,* 413-423.
14. Morita, A.; Motojima, K. *J. Weed Sci. Tech.* **1997,** *42(suppl. 1),* 164-165. (in Japanese)
15. Seitoh, F.; Tazawa, F.; Kawamura, M. *J. Weed Sci. Tech.* **1997,** *42(suppl.),* 166-167. (in Japanese)
16. Kohara, H.; Konno K.; Takekawa, M. *J. Weed Sci. Tech.* **1999,** *44,* 228-235. (in Japanese with English summary)
17. Yoshida, S.; Onodera, K.; Soeda, T.; Takeda, Y.; Sasaki, S.; Watanabe, H. *J. Weed Sci. Tech.* **1999,** *44(suppl.),* 70-71. (in Japanese)

18. Itoh, K.; Uchino, A; Watanabe, H. *J. Weed Sci. Tech.* **1998,** *43(suppl. 1),* 40-41. (in Japanese)
19. Hata, K.; Okuda, K.; Aoki M.; Kurimoto, H. *J. Weed Sci. Tech.* **1998,** *43(suppl. 1),* 28-29. (in Japanese)
20. Koarai, A.; Morita, H. *Kanto Zakken* **1999,** *10. (In Japanese)*
21. Aoki, M.; Kuramochi H.; Hata, K.; Otsuka, K. *J. Weed Sci. Tech.* **1998,** *43(suppl. 1),* 34-35. (in Japanese)
22. Sugimoto, M.; Ohashi, Y.; Kawase, K. *J. Kinki-Sakuiku-Kenkyu* **1999,** *44,* 43-48. (in Japanese)
23. Wang, G. X.; Watanabe, H.; Uchino A.; Itoh, K. *J. Weed Sci. Tech.* **1998,** *43(suppl. 1),* 38-39. (in Japanese)
24. Itoh, K.; Wang, G. X.; Shibaike H.; Matsuo, K. In *Proc. 17th Asian-Pacific Weed Science Society Conf.;* Bangkok, 1999; pp 537-543.
25. Gerwick, B. C.; Mireles, L. C.; Eilers, R. J. *Weed Technol.* **1993**, *7,* 519-524.
26. Uchino, A.; Watanabe, H.; Wang, G. X.; Itoh, K. *J. Weed Sci. Tech.* **2000,** *45,* 13-20 (in Japanese with English summary)
27. Matsumura, T.; Wang, G. X.; Goukon, K.; Itoh, K. In *Proc. 17th Asian-Pacific Weed Science Society Conf.;* Bangkok, 1999; pp 850-855.
28. Wang, G.X.; Watanabe, H.; Uchino, A.; Itoh K. *J. Weed Sci. Tech.* **1998,** *43(suppl. 1),* 42-43. (in Japanese)
29. Shibaike, H.; Uchino, A.; Itoh, K. In *1999 Brighton Conference-Weeds-Proceedings;* The British Crop Protection Council, Ed,; Brighton, UK, 1999; Vol. 1, pp197-202.
30. Uchino, A.; Watanabe, H. *J. Weed Sci. Tech.* **1999,** *44(suppl. 1),* 80-81. (in Japanese)
31. Shibuya, K.; Yoshioka, T.; Saitoh, S.: Yoshida, S.; Hashiba, T. *J. Weed Sci. Tech.* **1999,** *44(suppl. 1),* 72-73. (in Japanese)
32. Itoh, K.; Wang, G. X.; Uchino, A.; Tachibana, M. *J. Weed Sci. Tech.* **1997,** *42(suppl. 1),* 24-25. (in Japanese)
33. Watanabe, H.; Wang, G. X.; Uchino, A.; Itoh, K. *J. Weed Sci. Tech.* **1998,** *43(suppl. 1),* 46-47. (in Japanese)
34. Fukumoto, T.; Kagawa M.; Shirakura, S. *J. Weed Sci. Tech.* **1999,** *44(suppl. 1),* 90-91. (in Japanese)
35. Sugiura, K.; Watanabe, S.; Soeda, T. *J. Weed Sci. Tech.* **1999,** *44(suppl. 1),* 86-87. (in Japanese)

Chapter 12

Development of Herbicide-Resistant Upland Weeds in Japan

Yasuhiro Yogo

National Agricultural Research Center, Tsukuba, Ibaraki, 305–8666, Japan

In upland fields in Japan, resistant weeds were confirmed only on paraquat and simazine at the moment. Paraquat resistant *Erigeron philadelphicus* was first found in mulberry fields located along the Arakawa River in Fukiage/Saitama in 1980, and subsequently in orchard fields at several locations. The other Composite species, such as *E. canadensis*, *Conyza Sumatrensis* and *Yungia japonica* were also recognized as paraquat resistant in 1980's. In these fields, only paraquat had been used as a split application for several years continuously, and resistant biotypes were 50-100 times tolerant to susceptible biotypes. It was assumed that the single dominant gene or closely linked set of genes was responsible for the resistance. Simazine resistant *Poa annua* was also found in golf courses in Kansai district, where simazine had been applied for more than several years. This weed is cross-resistant to other triazine herbicides but not to 2,4-D. In case of simazine, the study was very limited in viewpoint of mode of resistance.

© 2002 American Chemical Society

Introduction

The herbicide resistant weeds in Japan were surveyed to local researchers, who conducted the herbicide registration trials by questionnaire in 1989 (1). Most of the herbicide resistant weeds were mainly Compositae family, such as *Erigeron philadelphicus* L., *E. canadensis* L. and *Conyza sumatrensis* (Retz.)Walker (*Table 1,2*). *E. philadelphicus* was distributed in Kanto district, central Japan, *E. canadensis* and *C. sumatrensis* were distributed central to southern Japan. Those resistant weeds appeared in 1984 to 1987, where paraquat were used 2-3 times per year in most cases. Therefore, those weeds should be resistant against paraquat, although the actual confirmation was rarely conducted on the resistance after the questionnaire. Very limited cases were recognized in triazine resistant weeds, although simazine and atrazine were registered in 1958 and 1966, respectively, and were widely used in upland fields in Japan. In this chapter, I listed up "herbicide resistant weeds", which were confirmed by field test treated with concerned herbicides (Stage 3 in *Table 3*) or comparative study on response to the herbicides between S and R biotypes/strains (Stage 4, *ditto*).

Paraquat Resistant Weeds

Before conducting the above questionnaire, paraquat resistant weeds were investigated by many researchers in Japan (*Table 3*). The resistant weeds were all Compositae family. Paraquat resistant *E. philadelphicus* was first found in mulberry fields at Fukiage city, Saitama prefecture in 1980 (2,3,4). The fields was located along the Arakawa River. *E. canadensis* (5), *Youngia japonica* (6) and *Conyza sumatrensis* (7) were also found in Saitama prefecture. *E. canadensis* (8) was also found in Osaka, however, paraquat resistant *E. philadelphicus* was not found there. All of them were found in mulberry/orchard fields or roadside. The resistant weeds were also observed in the other locations, from Tohoku to Kyushu, but not found in Hokkaido.

In the case of paraquat, the leaf disc test is reliable for confirming paraquat resistant weeds, therefore it is treated as well as Stage 4. The R biotype of *C. sumatreinsis* was ca. 1000 times resistant than S biotype in paraquat, although it was less resistant to diquat than paraquat (*Figure 1*) The degree of paraquat resistance, R/S ratio, was 50-100 or more in many cases (*Table 3, Figure 2*).

Table 1. Questionnaire results on location of herbicide resistant weeds

Weeds	Number of place with resistant weeds									
	A*	B	C	D	E	F	G	H	I	J
Grass	0	0	2	2	0	0	1	0	1	0
Compositae	3	23	31	28	9	17	28	10	12	2
Cruciferae	0	0	1	0	0	0	0	0	1	0
Total	3	23	34	30	9	17	29	10	14	2

*: A = Paddy, B = Field crop, C = Orchard, D = Mulberry, E = Grass land, F = Cultivation Abandons, G = Roadside, H = Garden, I = Public place, J = Others

Table 2. Questionnaire results on herbicide use and resistant weed occurence

Weeds	Number of place with resistant weeds						
	A*	B	C	D	E	F	G
Grass	4	1	1	0	0	1	0
Compositae	58	5	5	2	1	0	1
Cruciferae	1	0	0	0	0	0	0
Total	63	6	6	2	1	1	1

*: Herbicides; A = paraquat, B = glyphosate, C = simazine, D = glufosinate, E = bialaphos,

F = asulam, G = trifluralin

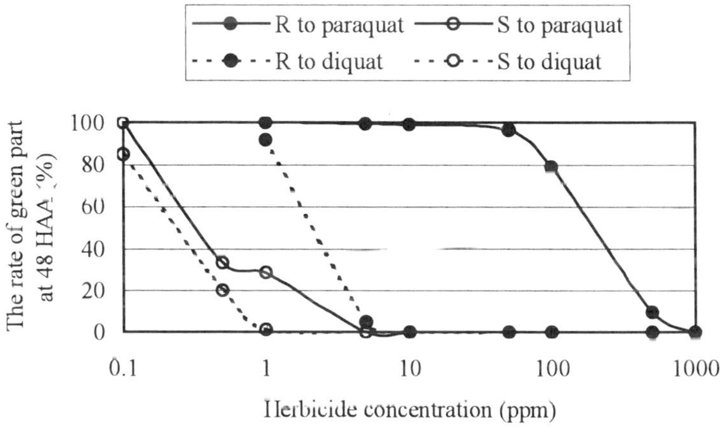

Figure 1. Effect of paraquat and diquat to leaf discs of Conyza sumatreinsis

Table 3. List of paraquat resistant weeds in upland fields, Japan

Weeds	Location	Place	Confirm. (year)*	Degree of Resistance	Ref.
Erigeron philadelphicus L. (6 strains)	Fukiage /Saitama	Mulberry	S2(1980) S3(1980) S4(1980-1) S5(1981-2)	50-100	2 3 4
Erigeron philadelphicus L.	Saitama	Mulberry	S5(1982)	50-100	9 10
Erigeron philadelphicus L.	Tsukuba /Ibaraki	Mulberry	S4(1984) S5(1984)	>100	-
Erigeron philadelphicus L.	Ibaraki	Mulberry	S4(1984) S5(1984)	100	11
Erigeron philadelphicus L.	Tohoku district	Orchard Roadside	S5(1984)		-
Erigeron canadensis L.	Osaka	Orchard	S2(1980) S3(1981)	1000	8 12
Erigeron canadensis L.	Kobe /Hyogo	Farmland	S5(1983)	1000	-
Erigeron canadensis L.	Saitama	Mulberry	S5(1986)	100	5 10

Youngia Japonica (L.)DC.	Saitama	Mulberry	S4(1986) S5(1986)	Max.>100	6
Conyza sumatrensis (Retz.)Walker	Kumagaya /Saitama	Mulberry	S4(1986) S5(1986)	>100	7
Graphalium pensylvanicum Willd.**	Kanagawa	Mulberry	S5?(1992)		12
Trigonotis peduncularis (Trevir.)Benth.**	Osaka	Orchard	S2(1986)		12
Mazus pumilus (Burm.f.)V.Steenis**	Osaka	Orchard	S2(1985)		12
Solanum nigrum L.**	Osaka	Orchard	S2(1986)		12

* : Stage of confirmation of herbicide resistant weeds

S1 Questionnaire only (no observation)

S2: Field observation

S3: Field test with related herbicides

S4: Comparative *in vivo* or *in vitro* study between S and R biotypes/strains

S5: Field survey of distribution

**: These were not treated as paraquat resistant weeds, here.

Figure 2. Dose response of resistant(lower) and susceptible (upper) biotype of E. philadelphicus to paraquat

Dosage were 0, 5, 10, 20, 40, 80 and 160 g a.i./ha (from left to right)

After confirming paraquat resistance, several investigations were done in the problem fields, concerning time of appearance, change in time course of distribution and so forth (2,3,5,6,7,9,10,13).

Time to Develop Paraquat Resistant Weeds

In Saitama prefecture, the change of weed distribution was surveyed in 1960's to 1980's (*Figure 3*). Compositae weeds, such as *E. philadelphicus* increased over time, although the other winter weeds, *Stellaria media* and *Ixeris stolonfera*, decreased. Summer weeds, such as *Digitaria ciliaris*, remained dominat weeds and *Calystegia japonica* was not changed dramatically. Minimum or none cultivation system was started in early 1960's, and paraquat was used in mid-late 1960's. The application area and number of application per year of paraquat were increased at least until mid. 1980's (10,13). In the above mentioned mulberry fields in Fukiage city, paraquat was used 2-3 times per year starting from 1969-70 , and based on the questionnaire (2). R biotypes were first found in 1980. The situations in the other cases were similar to *E. philadelphicus* in Fukiage on paraquat resistant weeds. Therefore, it took about 10 years for R biotypes to have appeared (*Table 4*).

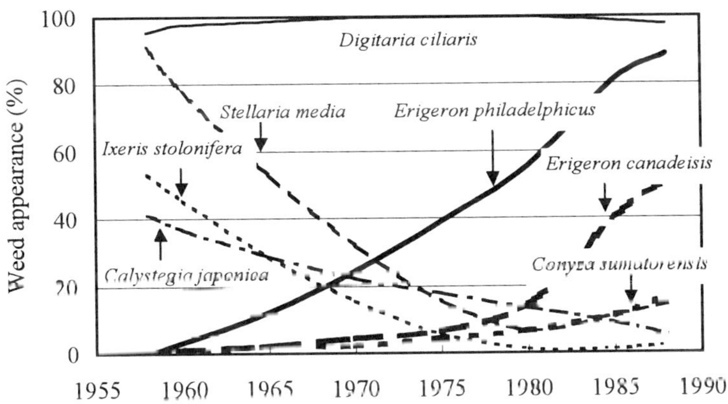

Figure 3. Change in weed appearance in Saitama, Japan

Table 4. Questionnaire on paraquat use and the resistant weeds

Paraquat					*Erigeron phiradelphicus* L.			
Start to use		Frequency/year			Initial infestation		R biotype first observed	
Year	%	No.	Past (%)	Current (%)	Year	%	Year	%
1969	26.3	1	2.6	0	1975	28.9	1979	5.3
1970	36.8	2	44.7	5.3	1970	39.5	1978	31.6
1971	7.9	3	44.7	44.7	1965	10.5	1977	28.9
1972	10.56	4	5.3	42.1	1960	10.5	1976	15.8
1973	2.7	5	2.7	7.9	1945<	2.7	1975<	7.9
1794	7.9				UC	7.9	UC	10.5
UC*	7.9							

*: UC = uncertain

The percentage here is to 38 mulberry growers who answered to the questionnaire out of 48, where paraquat resistant *E. philadelphicus* was observed in Fukiage/Saitama.

Distribution of Resistant Biotype

In many cases, R and S biotypes were mixed flora in the problem fields, as determined with the leaf disc test. The distribution rate of R biotype was not uniform in the problem fields. In mulberry fields along the Arakawa river in Fukiage, R biotype was dominant in the single center of the problem area, and S biotype was dominant in the surrounding fields (3). Interestingly, R biotype was not found in none-arable land or the abandoned mulberry field, where paraquat was not applied for 2-3 years in 1981. Distance between R biotype area and S-biotype area was less than 100 m. On the other hand, the opposite tendency was observed in the mulberry field in Fukaya/Saitama, where R type was dominant near road side (4). In a certain field in Kumagaya city, different distribution of R biotype among weed species was observed. It was supposed that *Y. japonica* invaded from north-east road side, although *E. philadelphicus* was from mulberry field on the north side and *E. canadensis* was dotted in the problem field (6). Paraquat resistant weeds might also appear at the same year in a different place, based on the macro mesh observation, such as city level. The cultivation methods in the fields, where R biotype was found, were minimum or none-tillage in many cases.

Hanioka (14) monitored the change of distribution of paraquat resistant *E. philadelphicus* from 1980 to 1998. In the 1980's, the number of fields with R biotype and the rate of R in monitored fields increased. In recent years, the R biotype can be observed in most of the surrounding fields, however, the R/S ratio was decreased (*Figure 4*).

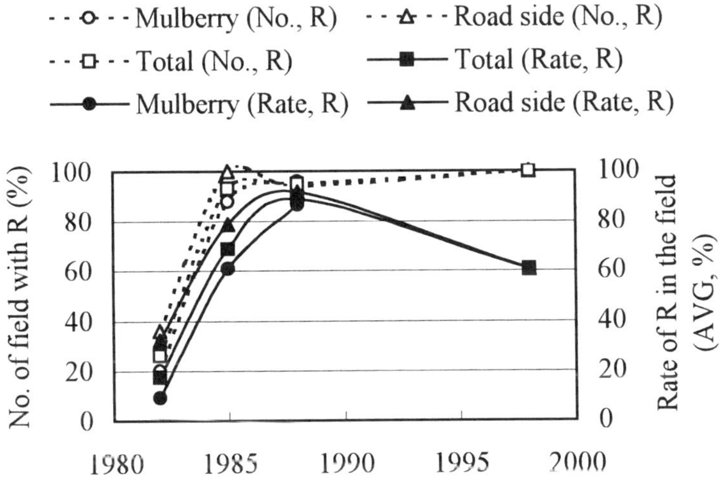

Figure 4. Change in time course of distribution of paraquat registant Erigeron philadelphicus in Saitama/Japan

Simazine Resistant Weed

Simazine resistance was only found in *Poa annua* L. in golf course of Kansai district, south-western Japan (*Table 5*). It was first found in 1984, where simazine use began in ca. 1975. Ohnishi (5) collected several biotypes of *P. annua*, and applied cyanazine. It was clear that there are different degrees of resistance among biotypes, i.e. R/S ratios of R biotypes ranged from 50-100 (*Figure 5*).

In the places where simazine resistant weeds were observed, simazine had been used for more than 7-8 years, starting from ca. 1975. Firstly, simazine was applied at ca. 1 kg a.i./ha, however, due to insufficient activity, the dosage was increased up to 4 kg a.i./ha year by year. These places were, of course, almost none-tillage, and might have appeared at the same year in the different place, as well as paraquat. In Takarazuka GC, R biotype was not found in green or tea ground, where simazine was not used, even though R biotype was found in fairway or rough near those places. The R population in the fairway and rough was reduced after simazine use was stopped.

Figure 5. Response of biotypes of Poa annua to cyanazine

untreated (upper) and treated (lower, at 3kg a.i./ha)

Table 5. List of simazine resistant weeds in upland field, Japan

Weeds	Location	Place	Confirm. (year)*	Degree of Resistance	Ref.
Poa annua	Kyoto Takarazuka	Golf course	S1(1983) S4(1984)	>256	15

* : Stage of confirmation of herbicide resistant weeds, see Table 3

Table 6. Distribution of simazine resistant weeds in Japan

Place	Detail	Simazine use	Emergence of R (%)
Takarazuka GC	Tee ground	+	1.0
	Fairway	+	5.0
	Rough	+	8.0
	Approach	-	0.0
	Green	-	0.0
Kyoto GC	Tee ground	+	76.9
	Fairway	+	29.4
	Rough	+	69.6
	Approach	+	100.0
	Green	-	13.5
Other place	Paddy	-	0.0
	Upland	-	0.0

Mode of Paraquat Resistant Weeds

Detoxification

The mode of paraquat resistance was studied using *E. philadelphicus*. Super oxide dismutase (SOD). In R biotype SOD from leaves was two to four times higher than S biotype, although no difference was observed in the other cases (16). Peroxidase (POD) activity in R biotype was higher than S biotype, and a different isozyme pattern was observed between R and S biotypes by using isoelectric focusing method (16). The R biotype was also resistant to $KClO_3$, a active oxygen source. This result confirms that POD has an important role in paraquat resistance. It was also clear that paraquat resistance disappeared when KCN, which is an inhibitor of SOD and POD, was applied in a mixture with paraquat (17).

Absorption and Translocation

A unique test was designed as shown in *Figure 6*. In B, the color of leaf discs of S biotype was changed at 2×10^{-4} M or more, although the leaf discs were dead even at 4×10^{-6} M in D. The leaf discs of R biotypes in A and C survived until 2×10^{-3} M (18). ^{14}C-paraquat studies with excised leaves or leaf discs showed no difference in absorption between R and S biotypes (19). The R biotype was resistant in leaf discs and callus, however, it lost resistance in protoplasts. These results and auto-radiographs indicate that the R biotype reduced the penetration of paraquat in comparison to the S biotype. Also resistance was reduced to moderate levels by pealing the leaf surface layer (18). Aging also led the resistance. Consequently, R biotype might absorb paraquat less than S biotype.

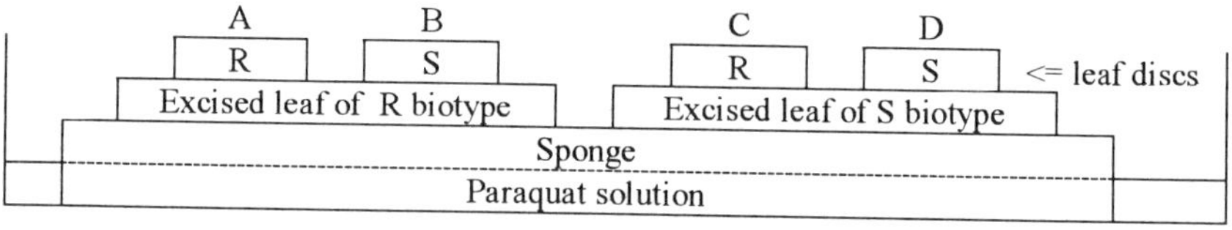

Figure 6. Effect of intermediate leaves on susceptibility of leaf discs of R and S biotypes of E. philadelphicus

Paraquat is known as a contact herbicide, however, it is also known that it will move in the dark, because the activity is higher in the evening application than in the morning application. A ^{14}C-paraquat study with excised leaves of *E. philadelphicus* and *E. canadensis* in the dark showed that paraquat moved more rapidly in S biotype than R biotype (5).

Competitiveness

Most of the studies supported that S biotype was more competitive than R biotype of *Erigeron spp.* without the paraquat stress based on photosynthesis, respiration, seed size, germination rate, growth under shading condition, flowering, mixed cultivation, de Wit method, and so forth (21,22,23,24,25,26). On the other hand, no clear difference in competitiveness was observed between R and S biotype of *Erigeron spp.*, based on the limited results (27,28).

Cross Resistant in Paraquat

Paraquat resistant weeds are resistant to dichloride and dimethylsulfate of paraquat and diquat. On the other hand, they were susceptible to other chemicals which have different modes of action from paraquat, such as ametryn, bentazone, bialaphos, bromacil, DCMU, glufosinate, glyphosate, MCP, 2,4-D (2,3,10).

Inheritance

F_1 progenies of *E. philadelphicus* between paraquat resistant and susceptible biotypes showed different segregation ratio of R/S, such as SxS = 0:1, SxR and RxS = 1:0 or 1:1, and RxR = 3:1 or 1:0 (*Table 7*). In addition, segregation ratio on R/S was all 1:1 in backcross test between paraquat resistant F_1 progenies and susceptible biotype (*Table 8*). It was concluded that paraquat resistance of *E. philadelphicus* was dominated by single dominant gene or closely linked set of genes (3,21,29). In case of *E. canadensis*, no clear segregation ratio was observed, although all of R and S biotype was clearly separated in viewpoint of the response to paraquat. Intermediate resistance was found in *Y. japonica*, although no detailed investigation was conducted. These weeds have different reproduction styles. *E. philadelphicus* has very low self-pollination and apomixes ratio, although those in *E. canadensis* and *Y. japonica* are high. It was supposed that the high out crossing ability of *E. philadelphicus* may lead to a high risk of R-biotype expansion not only by seeds but also by pollen in relation to single dominant paraquat resistant gene or closely linked set of genes.

Table 7. Segregation ratio between paraquat resistant and susceptible biotype of E. philadelphicus in F_1 progenies

female x male	Number of plants*			Expected ratio**
	Total	Survived	Dead	
Susceptible x Susceptible	502	0	502	0:1
	560	1	559	0:1
Susceptible x Resistance	147	147	0	1:0
	130	67	63	1:1
	203	94	109	1:1
Resistance x Susceptible	138	138	0	1:0
	115	56	59	1:1
	201	101	100	1:1
Resistance x Resistance	79	63	16	3:1
	136	136	0	1.0
	1122	1117	5	1:0
	94	94	0	1:0

* : Paraquat at 0.5 kg a.i./ha was applied at 1.5 leaf stage.

**: R/S ratio as a single dominant gene

Table 8. Segregation ratio of backcross test between paraquat resistant F_1 progenies and susceptible biotype of E. philadelphicus

female x male	Number of plants*			Expected ratio**
	Total	Survived	Dead	
Resistance (SxR) x Susceptible	100	43	57	1:1
Resistance (RxS) x Susceptible	989	491	498	1:1
Susceptible x Resistance (SxR)	345	164	181	1:1
Susceptible x Resistance (RxS)	305	147	158	1:1

* : See *Table 7*.
**: as a single dominant gene

Mode of Simazine Resistant Weed

Competitiveness

S biotype of *P. annua* is more competitive than R biotype with mixed cultivation. Flowering time of S biotype also was earlier than R biotype, although seed dormancy of R biotype was deeper than S biotype.

Cross Resistant in Simazine

Simazine resistant *P. annua* was also clearly resistant to the other triazines (e.g. atrazine, cyanazine) and diazines (e.g. chloridazon, lenacil), which have the same site of action as simazine. On the other hand, R biotype was moderately resistant to chlorophthalim, and susceptible to asulam, bethrodine, bifenox, butamifos, bithiopyr, flazasulfuron, orthobencarb, pendimethalin, prodiamine, propyzamide, pyributicarb and 2,4-D, which have different site of action from simazine(15).

Summary

Weeds resistant to bipyridilium herbicides, such as paraquat, were found in 18 dicots and 7 monocots internationally, and triazine resistant weeds were found in 42 dicots and 19 monocots. In Japan, 4 bipyridilium herbicide resistant weeds were found and confirmed, and all of them were Compositae family. *Poa annua* was only found as a triazine resistant weed, and no triazine resistant dicots were found until now. Paraquat was used as a single product or in mixture with diquat 2-3 times per year. Gramineae family, *Amarantus* spp. and *Solanum* spp., are generally observed in the mulberry/orchard fields and roadsides. The resistant weeds from these families were found in other countries. We cannot determine the reason why paraquat resistant weeds were found only in Compositae family in Japan. In golf courses in Japan, simazine is used as a grass killer in mixtures with dicots killers, although simazine itself is a dicots killer in general. We estimate that *P. annua* is susceptible only to simazine, but is resistant to the mixing partners. On the other hand, dicots were susceptible not only to simazine but also to the partners. Therefore the situation there is a continuous single application for *P. annua*, and this might be the reason for resistance.

The following items were common in the places where herbicide resistant weeds were found in upland field in Japan.

Continuous use of paraquat
Minimum/none-tillage field
Several locations at the similar timing
Expansion from single point/edge in one field

These conditions were generally observed in many other cases of herbicide resistant weeds in the world, as well as insecticide/fungicide resistance. Based on these results, paraquat and simazine select for R biotype to compete S biotype, and the R biotype may expand by seeds or propagation/hybridization. R-biotype exist at low levels prior to selection. Since upland farming size per farmer in Japan is less than 1 ha in general, farmers cultivate field and rotate crops in many cases within or between small patches. It might be the reason why herbicide resistant weeds have not appeared in such places.

We have no clear conclusion on the mode of resistance of paraquat. Detoxification by SOD and/or POD, limited absorption and/or uneven distribution of paraquat may contribute to the resistance of *E. philadelphicus*. The paraquat resistance might be realized with different factors, or the factor is different among biotypes and/or species. But it was clear that paraquat resistance was dominated by single dominant gene in *E. philadelphicus*. Therefore, it is easy to understand the detoxifying enzyme had a important role for the resistance.

In case of simazine, only the limited results on the resistant *P. annua* are available in Japan. However, since there is a wide range of the degree of resistance, not only modification of D1 protein but another factors may contribute to this mode of resistance.

There are two goals in the study in the herbicide resistant weeds, one is to solve problem in the fields, and the other is to clarify mode of resistance. As I mentioned before, herbicide resistant weeds problem are not a serious problem in upland fields in Japan, and we have certain solution for these problems, such as alternative herbicides, at the moment. Therefore, it is not urgent to monitor

ARTHUR LAKES LIBRARY
COLORADO SCHOOL OF MINES
GOLDEN, CO 80401

resistant weeds, or find and establish appropriate solutions. On the other hand, there is still a lot to be clarified, such as mode of appearance of resistant weeds in Japan.

Acknowledgement

I appreciate to Dr. Y. Watanabe, Dr. Y. Hanioka, Dr. M. Satoh, Dr. K. Itoh, Dr. H. Asano, Mr. T. Ohnishi and their colleagues to provide all data here and for their kind advice.

References

1. M. Sato, K. Itoh, Y. Usami and H. Koizumi; *Weed Research, Japan*, **1990**, *35*, 377-384 (in Japanese)
2. Y. Watanabe, T. Honda, K.Itoh and M. Miyahara; *Weed Research*, Japan **1982**, *27*, 49-53
3. K. Itoh; *JARQ*, **1988**, *22*:85-90
4. Y. Hanioka; *Weed Research, Japan*, **1983**, *28*, 213-214 (in Japanese)
5. Y. Hanioka; *Weed Research, Japan*, **1991**, *36*, 298-300 (in Japanese)
6. Y. Hanioka; *Weed Research, Japan*, **1989**, *34*, 163-168 (in Japanese)
7. Y. Hanioka; *Weed Research, Japan*, **1989**, *34*, 210-214 (in Japanese)
8. A. Kato and Y. Okuda; *Weed Research, Japan*, **1983**, *28*, 54-56 (in Japanese)
9. Y. Hanioka; *Weed Research, Japan*, **1989**, *34*, 215-221 (in Japanese)
10. Y. Hanioka; *Weed Research, Japan*, **1993**, *38*, 76-81 (in Japanese)
11. Y. Usami, H. Koizumi, H. Saka and M. Sato; *Weed Research, Japan*, **1989**, *34*, 57-61 (in Japanese)
12. Kato; *Osaka Nougyou*, **1987**, *24*, 16-22 (in Japanese)
13. Y. Hanioka; *Bull. Saitama-ken Sericultural Experiment Station*, **1992**, *40*, 1-78 (in Japanese)
14. Y. Hanioka; *Proceeding of 2nd Research Seminar of Division of Vegetation Science in NIAES*, **1998**, 15-23 (in Japanese)
15. T. Ohnishi and Y. Kirimura; *Bull. Hyogo Pre. Agri. Inst. (Agriculture)*, **1995**, *43*, 9-16 (in Japanese with English summary)
16. H. Asano; *Weed Research, Japan*, **1990**, *35*, 109-115 (in Japanese with English summary)
17. H. Asano Ph.D. thesis; Nihon University, Tokyo, Japan, **1990**, (in Japanese with English summary)
18. H. Asano; *Weed Research, Japan*, **1990**, *35*, 13-19 (in Japanese with English summary)

19. H. Asano; *Weed Research, Japan*, **1990**, *35*, 20-24 (in Japanese with English summary)
20. Y. Tanaka, H. Chisaka and H. Saka; *Physiol. Plant.*, **1986**, *66*, 605-608
21. K. Itoh and S. Matsunaka; In *Biological Approaches and Evolutinary Trends in Plants*; S. Kawano Ed., Academic Press, London, UK, 1990, pp 33-49
22. K. Itoh and M. Miyahara; *Weed Research, Japan*, **1983**, *28(Suppl.)*, 187-189 (in Japanese)
23. S. Saka, T. Osada and H. Chisaka; *Weed Research, Japan*, **1983**, *28(Suppl.)*, 193-194 (in Japanese)
24. T. Hirata and S. Matsunaka; *Weed Research, Japan*, **1985**, *30(Suppl.)*, 127-128 (in Japanese)
25. H. Asano; *Weed Research, Japan*, **1986**, *31(Suppl.)*, 83-84 (in Japanese)
26. S. Matsunaka and A. Moriyama; *Weed Research, Japan*, **1987**, *32(Suppl.)*, 141-142 (in Japanese)
27. M. Satoh, K. Usami and H. Koizumi; *Proceedings of 14th Asian-Pacific Weed Science Conference*, **1993**, 302-305
28. M. Satoh *et.al.* ; *Weed Research, Japan*, **1992**, *37*, 167-168 (in Japanese)
29. K. Itoh and M. Miyahara; *Weed Research, Japan*, **1984**, *29*, 301-307

Fungicide Resistance

Chapter 13

Mechanism of Resistance to Dicarboximides and Phenylpyrroles

Makoto Fujimura[1], Noriyuki Ochiai[2], Takayuki Motoyama[3], and Isamu Yamaguchi[3]

[1]Faculty of Life Science, University of Toyo, Itakura 374–0193, Japan
[2]Department of Engineering, University of Toyo, Kujirai 350–8585, Japan
[3]Microbial Toxicology Laboratory, Riken Institute, Hirosawa, 351–0198, Japan

The osmotic-sensitive *os-1, os-2, os4, and os-5* mutants of *Neurospora crassa* were found resistant not only to iprodione but also to fludioxonil. Stimulation of glycerol biosynthesis by these fungicides observed in the wild-type strain was not induced in the *os* mutant strains. The *os-1* mutant alleles were classified into two groups based on fungicide sensitivity and osmotic stress: Type I (highly resistant to fungicides but moderately sensitive to osmotic stress) and Type II (highly sensitive to osmotic stress but moderately resistant to fludioxonil). Sequence analysis of the *os-1* gene in Type I mutant NM233t revealed that the codon for amino acid position 308 was changed to a stop codon in the *os-1* gene, suggesting that an *os-1* product, which is predicted to be the osmosensor, histidine kinase, may be required for these fungicides to express their antifungal activity. Based on the characterizations of the *os* mutants in *N. crassa*, possible mechanisms of laboratory and field dicarboximide-resistance in *Botrytis cinerea* are also discussed.

 © 2002 American Chemical Society

Introduction

Dicarboximides, such as iprodione, procymidone, and vinclozolin, have been applied against *Botrytis cinerea* on a variety of crops, including grapes, vegetables, and ornamentals (1). Dicarboximides were primarily introduced in an attempt to combat resistance to benzimidazoles. However, within several years of their initial application, dicarboximide-resistant isolates of *B. cinerea* appeared on a variety of crops worldwide (1-4). Under laboratory conditions, dicarboximides easily select resistant strains in many kinds of fungi, including *B. cinerea* (4-7). In general, most dicarboximide-resistant laboratory mutants acquired high resistance to dicarboximide fungicides, and were hypersensitive to high osmotic pressure. However, high-level resistant strains have seldom been recovered from the field. Low-level resistant strains without an osmotic-sensitive phenotype only cause problems in disease control. Both types of dicarboximide-resistant isolates show cross-resistance to aromatic hydrocarbons, such as quintozene (PCNB) and tolclofos-methyl.

Phenylpyrroles, derivatives of pyrrolenitrin produced by several *Pseudomonas* species, were discovered as a new class of fungicides (8, 9). Field isolates of *B. cinerea* resistant to dicarboximides and aromatic hydrocarbons continue to be sensitive to phenylpyrroles. Interestingly, Leroux *et al.* demonstrated a cross-resistance to dicarboximides and phenylpyrroles in laboratory mutants of different fungi (10). This unexpected feature could hint at a similar mechanism of action. The antifungal spectrum of phenylpyrroles, such as fludioxonil and fenpiclonil, is similar to that of dicarboximides, although the phenylpyrroles have a rather wide antifungal spectrum and are used as seed fungicides to control *Fusarium*, *Tilletia*, and other seed-bone pathogens (8,9). Dicarboximides and phenylpyrroles induce similar morphological changes in germination tubes (10, 11) and their fungicidal toxicity is reversed by piperonyl butoxide and α-tocopherol in *B. cinerea* and other fungi (10). These fungicides stimulate glycerol synthesis in the mycelium of *Neurospora crassa* (11, 12). Despite many years of investigation, the mode of action of dicarboximides is not known. We have chosen the fungus *N. crassa* because of its well-characterized genetics and molecular biology to study the mode of action and mechanism of resistance in these fungicides.

<u>Cross-resistance between Iprodione and Fludioxonil in Osmotic-sensitive Mutants.</u> Sensitivity to fungicides (iprodione, PCNB, and fludioxonil) and osmotic stress in *B. cinerea* is shown in Table 1. Field isolate Bc-DLR4 was moderately resistant to iprodione and PCNB but sensitive to fludioxonil. In contrast, laboratory resistant isolate Bc-DHR3 showed a high degree of

resistance not only to iprodione and PCNB but also to fludioxonil. The growth of isolate Bc-DHR3 was inhibited on a medium containing 4% NaCl. These observations indicate discrepancies between field resistance and laboratory resistance and confirm previous descriptions (10, 13, 14).

Table I Sensitivity to fungicides and osmotic stress of dicarboximide resistant isolates in *B. cinerea*

Strain	Fungicide sensitivity (μg/ml)[1] Iprodione	PCNB	Fludioxonil	Osmotic sensitivity[2] 4% NaCl
Bc-DS5 (wild-type)	0.3	2.8	0.003	20
Bc- DLR4 (field-R)	3.6	21	0.005	33
Bc- DHR3 (lab-R)	>100	>100	>100	94

[1] Concentrations to inhibit radial growth at 50% level

[2] % of growth inhibition on medium containing NaCl

The lack of cross-resistance between dicarboximides and phenylpyrroles in filed isolates of *B. cinerea* (Table I) and selective inhibition of protein kinase (PK III) by phenylpyrroles in the wild-type strain of *N. crassa* suggest that the target sites of the dicarboximides and phenylpyrroles are different (12). In *N. crassa*, osmotic sensitivity is determined by at least six genes; *os-1*, *os-2*, *os-4*, *os-5*, *cut*, and *sor(T9)* (15-17). Among the osmotic-sensitive mutants, the *os* mutants are resistant to dicarboximides and aromatic hydrocarbons, whereas *cut* and *sor(T9)* are not (18-20). We examined the sensitivity to fludioxonil in these osmotic-sensitive mutants to investigate cross-resistance between dicarboximides and phenylpyrroles (Table II). A clear cross-resistance among iprodione, quintozene, and fludioxonil was observed in all *os* mutant strains. The mutants with *os-2*, *os-4*, and *os-5* were highly resistant to not only dicarboximides and aromatic hydrocarbons but also to fludioxonil; they were capable of growth even in the presence of 100 μg/ml of fludioxonil. Another osmotic-sensitive *cut* strain, which was sensitive to dicarboximides and aromatic hydrocarbons, was as sensitive as the wild-type strain to fludioxonil. Our results indicated that at least four different gene mutations, *os-1*, *os-2*, *os-4*, and *os-5*, confer cross-resistance to both iprodione and fludioxonil without exception.

Table II Fungicide resistance and osmotic sensitivity of osmotic-sensitive mutants in *Neurospora crassa*

	Fungicide sensitivity (μg/ml)[1)]			Osmotic sensitivity[2)]	
Strain	Iprodione	PCNB	Fludioxonil	4% NaCl	1% NaCl
Wild-type	1.0	0.60	0.005	54	0
os-1					
NM233t	>100	>100	>100	88	0
Y256M209	>100	>100	>100	93	40
M155-1	4.6	4.4	0.021	100	91
M16	14	>100	0.087	100	77
P3282	6.1	>100	0.038	100	90
P5990	>100	>100	1.6	98	90
os-2 (ALS10)	>100	>100	>100	97	14
os-4 (Y256223)	>100	>100	>100	97	16
os-5 (1638)	>100	>100	>100	97	14
cut (2386)	1.1	0.40	0.006	92	19

[1)] Concentration to inhibit radial growth at 50% level

[2)] % of growth inhibition on medium containing NaCl

Classification of *os-1* Mutants. Each *os-1* mutant strain of *N. crassa* was obviously more resistant to these fungicides and more sensitive to osmotic stress than the wild-type strain. However, the *os-1* mutant strains could be separated into two groups based on fludioxonil sensitivity and osmotic sensitivity (Table II). The first group (Type I), represented by strains NM233(t) and Y256M209, showed a high resistance to fludioxonil, iprodione, and quintozene. They were not osmotically sensitive to 0.5% NaCl although they were sensitive to 4% NaCl, the moderately osmotic-sensitive phenotype. Conversely, the second group (Type II), represented by strains M155-1, M16, P3282, and P5990, showed a moderate resistance to fludioxonil but a high sensitivity to osmotic stress. The growth of Type II mutants was inhibited even on a medium containing 0.5% NaCl. Interestingly, the Type I mutants were more resistant to fludioxonil but less sensitive to osmotic stress than were the Type II mutants, suggesting that the level of resistance is not parallel to that of osmotic sensitivity in the *os-1* mutant alleles. The level of sensitivity to iprodione, PCNB, and fludioxonil in the Type II mutants was different to each other. Thus, the *os-1* mutations are polymorphic in their sensitivities to fungicide and osmotic stress.

Abnormal Accumulation of Glycerol Induced by Iprodione and Fludioxonil. Biochemical studies revealed that phenylpyrroles inhibit the transport of monosaccharides in *Fusarium sulphureum* (37, 38). They also stimulate the intracellular accumulation of neutral polyols, such as glycerol and mannitol. Induction of glycerol synthesis was observed in the wild-type strain Bc-DS5 of *B. cinerea* (Fig. 1a) by the treatment of iprodione even at 1 μg/ml. Field isolates Bc-LR4 did not accumulate glycerol on the medium containing iprodione (1 μg/ml), but the stimulation of glycerol synthesis occurred at a high concentration of iprodione (25 μg/ml). In contrast, iprodione did not stimulate glycerol synthesis in laboratory isolates Bc-DHR3 even at a concentration of 25 μg/ml. These results suggest that iprodione sensitivity is correlated with the stimulation of glycerol synthesis by iprodione.

In *N. crassa*, significant glycerol accumulation was also induced in the wild-type strain by the treatment of iprodione and fludioxonil, as described by Pillonel *et al* (12). In addition, we reported that aromatic hydrocarbons such as PCNB and tolclofos-methyl also stimulated glycerol biosynthesis in the wild-type strain, suggesting that fungicides to which *os* mutants were resistant specifically caused the stimulation of glycerol biosynthesis (11). To elucidate the mechanism of resistance in the *os* mutants, the glycerol content recovered from mycelium grown in a medium containing fludioxonil and iprodione was determined (Fig. 1b). Neither fludioxonil nor iprodione significantly induced glycerol biosynthesis in highly resistant *os-2*, *os-4*, and *os-5* mutants. When a highly resistant *os-1* mutant strain (NM233(t), Type I) was treated with fludioxonil and iprodione, stimulation of glycerol biosynthesis was not observed. The *os-1* mutant strain (M16, Type II), which is moderately resistant to fludioxonil and iprodione, accumulated the same level of glycerol as the wild-type strain by fludioxonil (10 μg/ml), but less glycerol by iprodione (10 μg/ml). Thus, the reduced glycerol accumulations in the *os* mutant strains appears to correlate well with their resistance levels to fungicides, implying that the fungicidal activity of these compounds might be caused by the abnormal accumulation of glycerol.

However, the fungicide-sensitive *cut* mutant strain did not accumulate glycerol by the treatment of fludioxonil and iprodione (Fig. 1b). The glycerol content of the *cut* mutant strain was at an undetectable level and significantly lower than that of the untreated wild-type strain. In addition, that the *cut* mutant strain did not accumulate glycerol by osmotic stress suggests a lack of glycerol biosynthetic ability in the *cut* mutant strain. Despite its lack of glycerol biosynthesis, the *cut* mutant strain was sensitive to fludioxonil and iprodione, suggesting that abnormal glycerol accumulation is not essential for fungal toxicity.

Response to osmotic stress, both in the wild-type strain and in *os* mutant strains caused the accumulation of substantial quantities of glycerol, although the total amounts of glycerol recovered from the *os* mutant mycelium were less than that recovered from the wild-type mycelium. For example, the *os-5* mycelium grown under high osmolarity contained approximately 60% less glycerol than the wild-type mycelium (Fig. 1b). Although fludioxonil and iprodione did not induce glycerol accumulation in the *os* mutant strains, they retained their ability of glycerol synthesis in response to osmotic stress. These results were observed in laboratory isolate Bc-HR3 of *B. cinerea* (Fig. 1a). Filamentous fungi might indicate another pathway for stimulation of glycerol synthesis with which fungicides do not interfere.

Isolation and Characterization of Revertant Mutants from *os-5* Mutant Strain. The *os-2*, *os-4*, and *os-5* mutant strains of *N. crassa* were highly resistant to phenylpyrroles, dicarboximides, and aromatic hydrocarbons, as shown above. To explain the relationship between osmotic sensitivity and fungicide resistance in *os-5* mutants, we isolated *os-5* revertant strains on the medium containing 6% NaCl. The revertant os-5R-18 restored osmotic sensitivity to the level of the wild type. Interestingly, however, the os-5R-18 strain was still resistant to fludioxonil and iprodione (Table III).

Table III Sensitivity to fungicides and osmotic stress of *os-5* revertant strains in *N. crassa* mutants

Strain	Fungicide sensitivity (µg/ml)[1] Iprodione	Fludioxonil	Osmotic sensitivity[2] 4%NaCl	2%NaCl	1%NaCl
Wild-type	1.5	0.006	27	0	0
os-5	>100	>100	100	89	41
os-5R-18 (*os-5/su(os-5)*)	>100	>100	27	0	0
su(os-5)	1.0	0.003	56	22	0

[1] Concentrations to inhibit radial growth at 50% level

[2] % of growth inhibition on medium containing NaCl

When the revertent os-5R-18 strain was crossed with the wild-type strain, the progenies divided into three groups according their phenotypes, wild-type (one half), *os-5* (one fourth), and revertant (one fourth), indicating that an additional mutation outside the *os-5* locus, named *su(os-5),* confers the phenotype of the revertant os-5R-18 strain (Table IV). We isolated the *su(os-5)* strain without the *os-5* mutation by tetrad analysis in this cross. The sensitivity to fludioxonil and osmotic stress in the *su(os-5)* mutant strain was very similar to

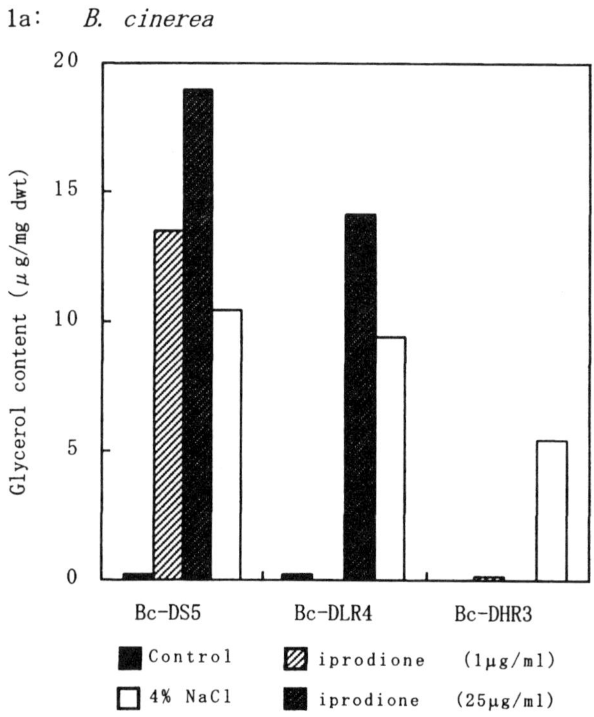

Fig. 1 Effect of iprodione and fludioxonil on the amount of glycerol in mutant strains in B. cinerea *(1a) and* N. crassa *(1b). Glycerol produced by the indicated strains was measured for mycelia grown for 4 hr in Vogel's medium N containing iprodione, fludioxonil or 4% NaCl.*

1b: *N. crassa*

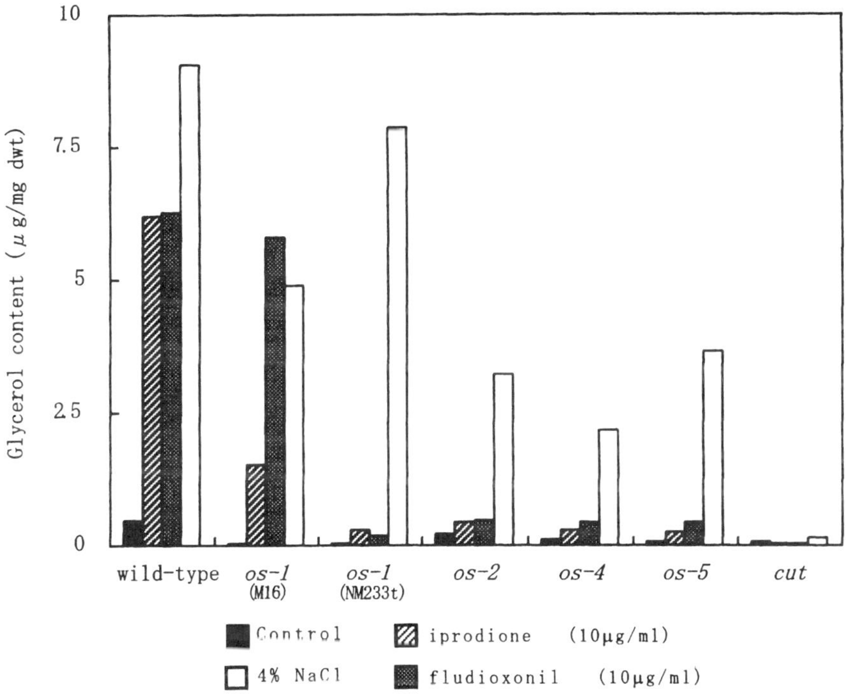

Figure 1. Continued.

that in the wild-type strain (Table III), suggesting that the mutation *su(os-5)* alone conferred neither increased tolerance to osmotic stress nor resistance to fludioxonil. The morphology and female fertility of the *su(os-5)* strain were also indistinguishable from those of the wild-type strain. When the *su(os-5)* mutant strain was crossed with the *os-1* (M16) and *os-2* (ALS10) strains, no progeny showed resistance to both fludioxonil and osmotic stress, suggesting that *su(os-5)* specifically reverts the osmotic sensitivity of the *os-5* mutation, but not that of the *os-1* and *os-2* mutations (Table IV).

Table IV Segregation pattern of progenies in crosses involving the revertant mutants of the *os-5* strain

Cross	Number of progenies FNSOSR[1]	FNROSS[2]	FNROSR[3]
os-5R-18 (*os-5/su(os-5)*) x			
Wild-type	43	22	19
os-5 (NM216o)	0	49	38
os-5R-18	0	0	55
***su(os-5)* x**			
Wild-type	81	0	0
os-5R-18	43	0	49
os-1 (M16)	41	(36)[4]	0
os-2 (ALS10)	46	51	0

[1] FNSOSR: Sensitive to fludioxonil (25μg/ml) but resistant to 4% NaCl.

[2] FNROSS: Resistant to fludioxonil (25μg/ml) but sensitive to 4% NaCl.

[3] FNROSR: Resistant to both fludioxonil (25μg/ml) and 4% NaCl.

[4] Number of progenies resistant to fludioxonil (0.1μg/ml) but sensitive to fludioxonil 25μg/ml.

The revertant *os-5/su(os-5)* strain restored glycerol synthesis induced by osmotic stress to the level of the wild-type strain, whereas the *os-5* strain accumulated approximately 60% less glycerol than the wild-type strain. Stimulation of glycerol synthesis was not induced in either the *os-5* or the *os-5/su(os-5)* strain in the medium with fludioxonil (39). The stimulation of glycerol synthesis by fludioxonil and osmotic stress in the *su(os-5)* strain was indistinguishable from that in the wild-type strain. This data suggests that osmotic tolerance of the revertant mutant strain is caused by the recovery of the ability to synthesize glycerol under high osmolarity. The suppression of the *os-5* mutation by *su(os-5)* was specific to glycerol synthesis in response to osmotic stress but it did not affect glycerol synthesis induced by fludioxonil.

Hyphal Morphology and Female Fertility in *os* Mutant Strains. The *os* mutant morphologies are characterized by short and close-cropped aerial hyphae, although morphologies are influenced by humidity (11, 17). Coincident with the aberrant morphology, *os* mutants have altered cell wall compositions (22, 23). Two groups of *os-1* allele mutants were also distinguishable by their morphologies. The Type II mutants produced significantly shorter and more close-cropped aerial hyphae than did the Type I mutants. In addition, the Type II mutants were female sterile due to a lack of protoperithecia formation, whereas the Type I mutants produced a few protoperithecia. In addition, the revertant *os-5/su(os-5)* restored osmotic sensitivity to the level of the wild-type strain to the same degree that it recovered hyphal morphology to that of the wild-type strain. This revertant strain showed normal protoperithecia formation and female fertility, while the parental *os-5* mutant strain was almost female sterile. The hyphal morphology and female fertility seemed to be linked with osmotic sensitivity rather than fungicide resistance (Table V).

Table V Hyphal morphology and female fertility in *os* mutants

	Fungicide resistance[1)]	Osmotic sensitivity[2)]	Hyphal morphology[3)]	Female fertility[4)]
Wild-type	S	R	-	+++
os-1 (NM233t)	HR	S	+	++
os-1 (M16)	R	SS	+++	-
os-5 (NM216o)	HR	S	++	+
os-5/su(os-5)	HR	R	-	+++

[1)]S: Sensitive to fludioxonil (0.1μg/ml), R: Resistant to fludioxonil (0.1μg/ml) but sensitive to fludioxonil (25μg/ml), HR: Resistant to fludioxonil (25μg/ml)

[2)]SS: Sensitive to 0.5%NaCl, S: Resistant to 0.5%NaCl but sensitive to 4%NaCl, R: resistant to 4%NaCl

[3)]- to +++: Normal hyphal morphology to significant abnormal hyphal morphology

[4)]+++ to -: Normal female fertility to female sterile

Identification of Amino-acid Substitution of *os-1* Mutant Strains. The *os-1* allele mutants were divided into two groups, as described above. To obtain structural information on the *os-1* gene product (predicted osmosensor histidine kinase), we cloned the *os-1* genes from *os-1* mutants NM233t and M16, using the wild-type *os-1* gene as a probe. The *os-1* gene cloned from the M16 strain (Type II), pOS1-M16, did not transform the *os-1* M16 strain to osmotic resistance, whereas the wild-type *os-1* gene complemented this mutant. We determined the nucleotide sequence of the coding region of the *os-1* (M16) gene using thirteen synthetic primers and detected a mutation responsible for amino acid substitution (40). A T-to-C transition occurred in the second base of codon 625, resulting in an amino acid change from leucine to proline at position 625 in Os1p. For the NM233t strain, we found a C-to-T transition in the first base of codon 308. This mutation changes the codon for ^{308}Gln to a stop codon.

Table VI Amino acid substitutions in two types of *os-1* mutant strain

	Amino acid substitution	Fungicide resistance[1)]	Osmotic sensitivity[2)]
Wild type		S	R
os-1 (NM233t)	^{308}Gln to Stop	HR	S
os-1 (M16)	^{625}Leu to Pro	R	SS

[1)]S: Sensitive to fludioxonil (0.1μg/ml), R: Resistant to fludioxonil (0.1μg/ml) but sensitive to fludioxonil (25μg/ml), HR: Resistant to fludioxonil (25μg/ml)

[2)]SS: Sensitive to 0.5%NaCl, S: Resistant to 0.5%NaCl but sensitive to 4%NaCl, R: resistant to 4%NaCl

Discussion

The osmotically sensitive *os-1*, *os-2*, *os-4*, and *os-5* mutants were resistant to dicarboximides, aromatic hydrocarbons, and phenylpyrroles. The *os-1* gene product of *N. crassa* is predicted to be an osmosensor in an osmotic response (NOR) signal transduction cascade (24, 25), similar to the HOG pathway in yeast (26, 27). The components of the NOR pathway are encoded by other osmotic loci such as *os-2* and *os-5* (33). Pillonel *et al.* reported that a protein kinase (PK-III) possibly involved in the regulation of the glycerol biosynthesis was inhibited by phenylpyrroles in *N. crassa* (12). In addition, Orth *et al.*

reported a c-AMP-dependent serine (threonine) protein kinase that confers vinclozolin resistance of *Ustilago maydis* (28). These findings suggest that dicarboximides and phenylpyrroles interfere with osmotic signal transduction.

The Os1p sequence is homologous with both the kinase and response regulator modules of two-component signal protein. It is novel, however, in that the N-terminal end of the polypeptide contains unique tandem repeats of 90 amino acids (24, 25). We cloned the *os-1* gene of strain M16 (Type II) and identified a single amino-acid substitution, ^{625}Lue to Pro. This amino-acid change is located at the end of the fifth repeat of 90 amino acids in Os1p. The growth of the M16 strain was highly inhibited on the medium containing 0.5% NaCl, while this strain had low resistance to fludioxonil. Thus, the structure of tandem repeats of 90 amino acids in Os1p, which is predicted to form a coiled-coil structure (29), is very important for osmotic adaptation. In the *os-1* NM233t strain, an amino acid at position 308 in Os1p was changed into a stop codon. As predicted, the *os-1* (NM233t) gene product lacks both a histidine kinase domain and a responsible regulator domain, meaning Os1p(NM233t) must be inactive. The inactivation of Os1p results in a high resistance to dicarboximides and phenylpyrroles. Therefore, dicarboximides and phenylpyrroles use the function of Os1p to express their antifungal activity. Interestingly, the strain M16 (^{625}Lue to Pro) was more sensitive to osmotic stress than the strain NM233t (stop codon at position 308).

We isolated a new mutation, *su(os-5)*, a predicted component of the NOR pathway. *Su(os-5)* suppressed the osmotic sensitivity of *os-5* mutants, although *su(os-5)* alone conferred neither increased tolerance to osmotic stress nor resistance to dicarboximides and phenylpyrroles. In addition, *su(os-5)* specifically suppressed the osmotic sensitivity of the *os-5* mutation, but not that of the *os-1* and *os-2* mutations. Interestingly, this mutation did not affect the fungicide-resistant phenotype of the *os-5* mutants. Recently, an *os-5* gene has been cloned and revealed to be a *HOG1* cognate (33). The *su(os-5)* gene product is predicted to interact with the *os-5* gene product (Os5p) and might recover the functions of Os5p resulting in a normal response to osmotic stress. However, suppression by *su(os-5)* did not affect the fungicide resistance of the *os-5* mutant strain.

It should be noted that the *os* mutants showed not only osmotic sensitivity and fungicide resistance but also abnormal hyphal morphology and reduced female fertility. As shown in Table V, osmotic sensitivity is linked with hyphal morphology and female sterility but not with fungicide resistance in both the *os-1* and *os-5* mutants.

Together with our findings, fungicide resistance might be determined in a manner different from osmotic sensitivity in the NOR pathway. Apparently, there are two functions in the NOR pathway, one for osmotic adaptation and one for some other function. It is known that yeast cells use the same signaling proteins (Ste20p, Ste11p, Ste7p, and Ste12p) in two different pathways, which receive different input signals and generate different outputs (34). Recently, *COS1* has been cloned from the opportunistic pathogen *Candida albicans* (30-32). The overall domain organization is conserved between Os1p and Cos1p, including the large N-terminal domain that contains the 90 amino acid repeats. The deletion mutant of the *COS1* gene in *C. albicans* exhibits defective hyphal formation, whereas this mutant is still resistant to high osmolarity (30). Os1p and Cos1p may be involved in some aspect of hyphal morphogenesis. The *os* mutant is defective in hyphal formation on a medium containing 4% NaCl (11). It is known that the *os-1* mutant forms protoplasts under special conditions (35). In addition, when grown on media containing iprodione and fludioxonil, wild-type conidia burst without germ tubes in the wild-type strain of *B. cinerea* and *N. crassa* (11, 21). Dicarboximides and phenylpyrroles may interfere with signaling for hyphal formation in the NOR pathway.

In *B. cinerea,* the phenotype of field isolates resistant to dicarboximides is different from that of laboratory mutants. The discrepancy between field and laboratory resistance implies that the resistance mechanism of the field isolates differs from that of the laboratory isolates. Nevertheless, Feretra *et al.* have isolated phenylpyrrole-resistant mutants *in vitro* and characterized them genetically (13). Their results suggest that the *Daf1* locus is responsible for both the field resistance and laboratory resistance phenotypes. The *Daf1* gene is polymorphic; at least five classes of alleles have been identified based on their effects on phenotype response to dicarboximides and phenylpyrroles in *B. cinerea.* Similar results were published by Hilber *et al* (14). We have a working hypothesis that *Daf1,* which confers field dicarboximide resistance in *B. cinerea,* might be an *os-1* gene homologue. Together with the *Daf1* gene of *B. cinerea,* the *os-1* gene of *N. crassa* is highly polymorphic. In *N. crassa,* all the *os* mutants, except the *os-1* strains, are highly resistant to fungicides including fludioxonil, and their phenotypes are similar to each other. In contrast, some *os-1* allele mutants of *N. crassa* remain sensitive to fludioxonil. In addition, osmotic sensitivity was not directly linked with fungicide resistance in the *os-1* mutants. All *os-1* mutants are sensitive to osmotic stress because they have been isolated as osmotically sensitive mutants. Thus, it might be possible that a specific mutation within the *os-1* gene shows a field resistant phenotype. Furthermore, most mutations in laboratory resistant isolates in *B. cinerea* were mapped on the *Daf1* locus (13). Similarly, the *os-1* mutants were most easily obtained as dicarboximide-resistant mutants among the *os* mutants (36). The function of *Daf1*, a polymorphic gene, remains to be investigated to clarify our hypothesis.

Acknowledgments

We thank Dr. Christian Pillonel for his generous gift of fludioxonil and Dr. Claude Selitrennikoff for providing cosmid 12:4D. This work was supported in part by a Grant-in-Aid for Scientific Research (No. 1066052) from the Ministry of Education, Science, Sports, and Culture of Japan.

Literature Cited

1. Pommer, E.; Lorenz, G. In *Modern selective fungicides -Properties, applications, mechanisms of action*; H. Lyr, Ed.; Gustav Fischer Verlag: New York, 1995; pp 99-118.
2. Beever, R. E.; Brien, H. M. R. *New Zealand J. Agri. Res.* **1983**, *26*, 391-400.
3. Leroux, P.; Frits, R. In *Mode of action of antifungal agents*; A. Trinci and J. Ryley, Eds.; British Mycological Society: London, 1984; pp 207-237.
4. Lorenz, G. In *Fungicide resistance in North America*; C. J. Delp, Ed.; American Phytopathology Society: St. Paul, MN, 1988; pp 45-51.
5. Beever, R. *Trans. Br. Mycol. Soc.* **1983**, *80*, 327-331.
6. Leroux, P.; Frits, R.; Gredt, M. In *Systematische Fungizide und Antifungale Verbindungen*; H. Lyr and C. Polter, Eds.; Abh. Akad. d. Wiss. DDR, Akademie-Verlag: Berlin, 1983; pp 79-88.
7. Edlich, W.; Lyr, H. In *Modern selective fungicides -Properties, applications, mechanisms of action*; H. Lyr, Ed.; Gustav Fischer Verlag: New York, 1995; pp 119-131.
8. Nevill, D.; Nyfeler, R.; Sozzi, D. *Proc. Brighton Crop Prot. Conf. Pests Dis.* **1988**, *1*, 65-72.
9. Gehmann, K.; Nyfeler, R.; Leadbeater, A.; Nevill, D.; Sozzi, D. *Proc. Brighton Crop Prot. Conf. Pests Dis.* **1990**, *2*, 339-406.
10. Leroux, P.; Lanen, C.; Frits, R. *Pestic. Sci.* **1992**, *36*, 255-261.
11. Fujimura, M.; Ochiai, N.; Ichiishi, A.; Usami, R.; Horikoshi, K.; Ymaguchi, I. *J. Pestic. Sci.* **2000**, *25*.
12. Pillonel, C.; Meyer, T. *Pestic. Sci.* **1997**, *49*, 229-236.
13. Faretra, F.; Pollastra, S. *Mycol. Res.* **1993**, *97*, 620-624.
14. Hilber, U.; Schwinn, F.; Schuepp, H. *J. Phytopathology* **1995**, *134*, 423-428.
15. May, L. *Genetics* **1969**, *63*, 781-794.
16. Murayama, T.; Ishikawa, T. *J. Bacteriol* **1973**, *115*, 796-804.
17. Parkins, D.; Radford, A.; Newmeyer, D.; Bjorkmann, M. *Microbiol. Rev.* **1982**, *46*, 426-570.

18. Beever, R. E.; Byrde, R. J. W. In *Fungicide resistance in crop protection*; J. Dekker and S. G. Georgopoulos, Eds.; Pudoc: Wageningen, 1982; pp 101-117.
19. Grindle, M.; Temple, W. *Neurospora Newsl.* **1982**, *29*, 16-17.
20. Grindle, M.; Temple, W. *Trans. Br. Mycol. Soc.* **1985**, *84*, 369-372.
21. Hisada, Y.; Kawase, Y. *Ann. Phytopath. Soc. Jpn.* **1977**, *43*, 151-158.
22. Livingston, L. *J. Bacteriol.* **1969**, *99*, 85-90.
23. Scott, W. A. *Annu. Rev. Microbiol.* **1976**, *30*, 50-104.
24. Alex, L.; Borkovich, K.; Simon, M. *Proc. Natl. Acad. Sci. USA* **1996**, *93*, 3416-3421.
25. Schumacher, M.; Enderlin, C.; Selitrennikoff, C. *Curr. Microbiol.* **1997**, *34*, 340-347.
26. Posas, F.; Wurgler-Murphy, S.; Maeda, T.; Witten, E.; Thai, C.; Saito, H. *Cell* **1996**, *86*, 865-875.
27. Maeda, T.; Wurgeler-Murphy, S. M.; H, S. *Nature* **1994**, *369*, 242-245.
28. Orth, A.; Rzhetskaya, M.; Pell, E.; Tien, M. *Appl. Environ. Microbiol.* **1995**, *61*, 2341-2345.
29. Lupas, A.; Dyke, M. V.; Stock, J. *Science* **1991**, *252*, 1162-1164.
30. Alex, L.; Korch, C.; Selitrennikoff, C.; Simon, M. *Proc. Natl. Acad. Sci. USA* **1998**, *95*, 7069-7073.
31. Nagahashi, T.; Mio, T.; Ono, N.; Okabe, T.; Arisawa, M.; Bussey, H.; Okabe, H. *Microbiol.* **1998**, *144*, 425-432.
32. Srikantha, T.; Tsai, L.; Daniels, K.; Enger, L.; Highley, K.; Soll, D.. *Microbiol.* **1998**, *144*, 2715-2729.
33. Zhang, Y.; Lamm, R.; Pillonel, C.; Xu, J.; Lam, S. *Proceeding of 20th Fungal Genetic Conference at Asilomar* **1999**, 72.
34. Gustin, M. C.; Albertyn, J.; Alexander, M.; Davenport, K. *Microbiol. Mol. Biol. Rev.* **1998**, *64*, 1264-1300.
35. Selitrennikoff, C. P.; Lilley, B. L.; Zucker, R. *Exp. Mycol.* **1981**, *5*, 155-161
36. Grindle, M. *Trans. Br. Mycol. Soc.* **1984**, *82*, 635-643.
37. Jespers, A. B. K.; Davidse, L. C.; de Waad, M. A. *Pestic. Biochem. Physiol.* **1993**, *45*, 116-129.
38. Jespers, A. B. K.; de Waad, M. A. *Pestic. Sci.* **1995**, *44*, 167-175.
39. Fujimura, M.; Ochiai, N.; Ichiishi, A.; Usami, R.; Horikoshi, K.; Ymaguchi, I. *Pestic. Biochem. Physiol.* **2000**, *67*, 125-133.
40. Ochiai, N.; Fujimura, M.; Motoyama, T.; Ichiishi, A.; Usami, R.; Horikoshi, K.; Ymaguchi, I. *Pest Manage. Sci.* (In press)

Chapter 14

Resistance to Strobilurin Fungicides

Wolfram Köller[1], C. Avila-Adame[1], G. Olaya[2], and D. Zheng[1]

[1]**Department of Plant Pathology, Cornell University, New York State Agricultural Experiment Station, Geneva, NY 14456**
[2]**Western Research Center, Zeneca Ag Products, 1200 South 47th Street, Richmond, CA 94804–4610**

Strobilurin fungicides act as specific inhibitors of respiration by binding to cytochrome *b*, a membrane protein encoded by a mitochondrial gene. The target site of strobilurin-related inhibitors can be circumvented by the induction of alternative respiration. However, this particular mechanism of strobilurin resistance appears of moderate importance in the *in vivo* control of fungal plant pathogens. A potentially more significant mechanism of strobilurin resistance is a mutational change of the cytochrome *b* target site, which has been reported for *Saccharomyces cerevisiae* and several other organisms. In laboratory studies, a highly strobilurin-resistant mutant of the plant pathogen *Venturia inaequalis* was generated carrying a G143A exchange in the cytochrome *b* target site. In the presence of both a strobilurin-related fungicide and an inhibitor of alternative respiration, mutated mitochondria were selected to a high frequency, but sensitive mitochondria became dominant again as soon as cells were relieved from the strobilurin inhibitor implying a decreased fitness of mutated mitochondria. The risk of target site mutations will, thus, be determined by the relative fitness of mutated mitochondria in relation to the duration of selection pressure to which fungal cells will be exposed to during disease control.

© 2002 American Chemical Society

Introduction

The natural products myxothiazol, several oudemansins and the strobilurins have been characterized as potent inhibitors of respiration by binding to the Q_p center of cytochrome *b* as part of the ubiquinol-cytochrome *c* oxidoreductase (complex III) (*1, 2*). The relatively simple chemical structure of these antifungal compounds has provided the lead structure for a new class of commercial fungicides (*3*). Azoxystrobin and kresoxim-methyl were the first representatives of these strobilurin-related inhibitors (Figure 1), with metominostrobin and trifloxystrobin added more recently to this evolving list of new fungicides (4, *5*). In addition, the structurally different fungicide famoxadone was also described as an inhibitor of respiration by binding to the same center Q_p of cytochrome *b* (*6*).

The mode of action of strobilurin-related inhibitors was clarified with mitochondria isolated from mouse liver, but also from *Saccharomyces cerevisiae* and plants (*1-5*) suggesting a low degree of intrinsic inhibitor specificity. A favorable toxicological profile of respective fungicides has been related to secondary mechanisms such as rapid inhibitor detoxification (*7, 8*), but the relatively indiscriminate target affinity of strobilurin-related inhibitors explains the broad spectrum of disease control provided by these new fungicides (*4, 5*). Although systemic properties vary, all strobilurin fungicides exhibit slow movement within the cuticle, accompanied by more or less slow distribution throughout the plant. Because fungal spore germination is inhibited, deposits on plant surfaces will protect the host tissue from fungal infections. Both systemic distribution and pronounced anti-sporulant activities provide additional post-infection control of diseases (*4, 5*).

The highly specific action of strobilurin fungicides signals a considerable risk of resistance. As reported by the Fungicide Resistance Action Committee (http://www.gcpf.org/frac), relatively rapid development of practical resistance has been observed for powdery mildew of wheat and cucurbits. Predictive resistance risk assessments have been complicated by the fact that the cytochrome *b* target site is encoded by a mitochondrial gene (*9*). This difference from target sites of other fungicides has two implications: alleles of the gene are numerous within the large population of mitochondria, and mitochondria distinguished by a mutation of the cytochrome *b* would have to be selected first within individual phenotypes before such phenotypes would become fully resistant. Several mechanisms of fungal resistance to strobilurin-related inhibitors have been described and assessed, and their potential impact on the development of practical resistance will be the focus of this overview.

Mechanism of Strobilurin Resistance: Alternative Respiration

Impact of Alternative Respiration on Strobilurin Action

Under normal conditions of respiration, transfer of electrons through the mitochondrial electron transport chain involves cytochrome bc_1 (complex III).

As illustrated in Figure 2, plants and also many fungal organisms have evolved an alternative route of cyanide-resistant respiration utilizing alternative oxidase as the core element (*10, 11*). With alternative respiration active, electrons are transferred directly from the ubiquinol pool to oxygen, and the cytochrome bc_1 complex serving as the strobilurin target site will be circumvented. Electron flow through the alternative pathway is normally not accompanied by the generation of proton-motive force and, thus, is not coupled to ATP synthesis (Figure 2). In plants, alternative oxidases are encoded by a family of related nuclear genes. Disturbances of electron flow through the respiratory cytochrome chain will activate alternative respiration either by induction of alternative oxidase gene expression or by post-translational activation of constitutively expressed alternative oxidases (*10, 11*). For example, a post-translational mechanism was suggested to be active in the rescue of rice mitochondria from inhibition by the strobilurin fungicide metominostrobin (SSF 126) (*12*). Although not fully understood at present, the role of alternative respiration in plants is discussed as protection of mitochondria from oxidative stress and the partitioning of ATP and anabolic precursor synthesis between chloroplasts and mitochondria (*11, 13*).

In fungal organisms, both the properties and the biological roles of alternative respiration remain under current debate (*14*). Closely related alternative oxidase genes have been characterized for several fungal organisms including the rice blast pathogen *Magnaporthe grisea* (*Pyricularia oryzae*) (*14-16*). Respective genes were found as single-gene copies, but evidence provided for *Candida albicans* indicated the presence of an additional mechanism of cyanide-resistant respiration (*16*). Such departures from the alternative respiration scheme described for plants (Figure 2) were also described for the wheat 'take all' pathogen *Gaeumannomyces graminis* (*14*).

Regardless of several unanswered questions, it was noticed for a wild-type isolate of *M. grisea* that mycelium rapidly acquired resistance to the strobilurin-related inhibitor metominostrobin, and biochemical evidence indicated that the induction of alternative respiration was the likely cause of this resistance response (*17*). As illustrated in Figure 3 (unpublished results), this rescue response of *M. grisea* is very effective for both developmental stages of conidia germination and mycelial growth. Azoxystrobin exhibited only weak effects when applied alone, whereas inclusion of the alternative oxidase inhibitor salicylhydroxamic acid (SHAM) rendered both developmental stages highly and almost equally sensitive to the inhibitor. Such strong synergistic effects of an alternative oxidase inhibitor suggests that the circumvention mechanism illustrated in Figure 2 is highly effective in conferring resistance of fungi to strobilurin fungicides. Very similar evidence for the effectiveness of this circumvention mechanism has been presented for *Septoria tritici* (*18*), *Venturia inaequalis* (*19*) and *Botrytis cinerea* (*20*).

Direct proof for the involvement of alternative respiration in this mechanism of resistance has been provided for *M. grisea*. Expression of an alternative oxidase gene was rapidly induced upon treatment of mycelium with metominostrobin (*15*). Gene transcription was also induced in response to H_2O_2, a mechanism of induction relating to the generation of active oxygen in response to the inhibition of electron flow through the cytochrome pathway (*15*). Several

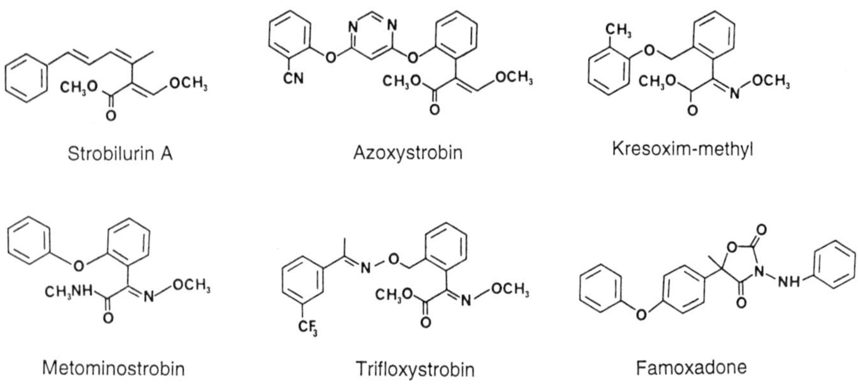

Figure 1. Structures of strobilurin-related fungicides.

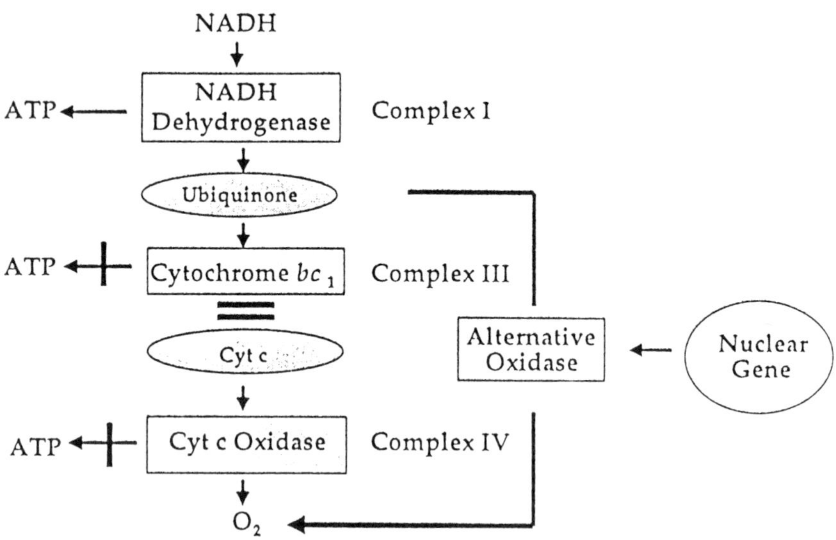

Figure 2. Scheme of alternative respiration. The strobilurin site of action is indicated by a double bar.

laboratory mutants more sensitive to metominostrobin lacked active alternative oxidase in mitochondria, and this inactivation of alternative respiration was accompanied by mutational changes of the alternative oxidase gene (*21, 22*). However, alternative respiration in other sensitized mutants was not affected (*21*).

Inhibitor responses of two mutants generated by restriction enzyme mediated insertion (REMI) (*23*) also exhibited a largely decreased effectiveness of the alternative respiration mechanism of resistance (Figure 4; unpublished results). However, this effect was not accompanied by an altered induction of the alternative oxidase gene. As indicated by the continued effect of SHAM, the effectiveness of alternative oxidase was only decreased but not necessarily abolished (Figure 4).

A diverse role of alternative respiration in the expression of strobilurin resistance was also described for a large number of baseline isolates of the apple scab fungus *V. inaequalis*. The sensitivities of germinating condia derived from 300 monoconidial isolates ranged from 0.003 to 0.14 μg/ml for kresoxim-methyl (*24*). At a low inhbitor dose, the synergistic effect of the alternative oxidase inhibitor SHAM was only apparent for 20% of the population. At a high inhibitor dose, however, SHAM acted as a strong synergist leading to full inhibition of conidia germination for all 300 isolates tested (*25*). The result indicated that the degree of cytochrome *b* saturation required for the induction of alternative respiration is a variable trait of individual phenotypes.

In summary, the induction of alternative respiration comprises an effective mechanism of rescuing fungal respiration from strobilurin action. Induction of this rescue pathway, however, is complex and requires the accumulation of active oxygen species to levels sufficiently high to induce expression of the alternative oxidase gene, an intact signal transduction pathway leading to gene induction and the import of functional alternative oxidase into the inner membranes of mitochondria. This cascade of events will depend on the interplay of multiple genes. For example, the expression of alternative respiration has been shown to be governed by a second major gene different from the alternative oxidase in *Neurospora crassa* (*26*). Within the population context of fungal pathogens, additional genes will likely contribute to the effectiveness of alternative respiration as a mechanism of strobilurin resistance.

Resistance Risks Inherent to Alternative Respiration

While the *in vitro* studies described above leave little doubt about the potential importance of alternative respiration as a mechanism of strobilurin resistance, the *in vivo* effectiveness of strobilurin-related fungicides appears to indicate that this mechanism is of minor importance under conditions of disease control. For example, the *in vitro* sensitivities of two isolates of *V. inaequalis* to kresoxim-methyl were drastically different (*19*). These differences were caused by the induction of alternative respiration, but they were not expressed in the *in vivo* control of apple scab (Table I). Very similar results have been reported for a laboratory mutant of *S. tritici* (*18*).

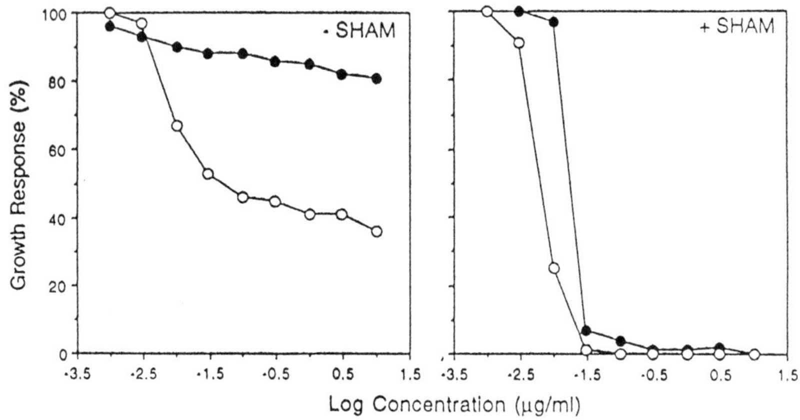

Figure 3. Response of Magnaporthe grisea *conidia (closed symbols) and mycelium (open symbols) to azoxystrobin in the absence and presence of SHAM at a concentration of (100 µg/ml).*

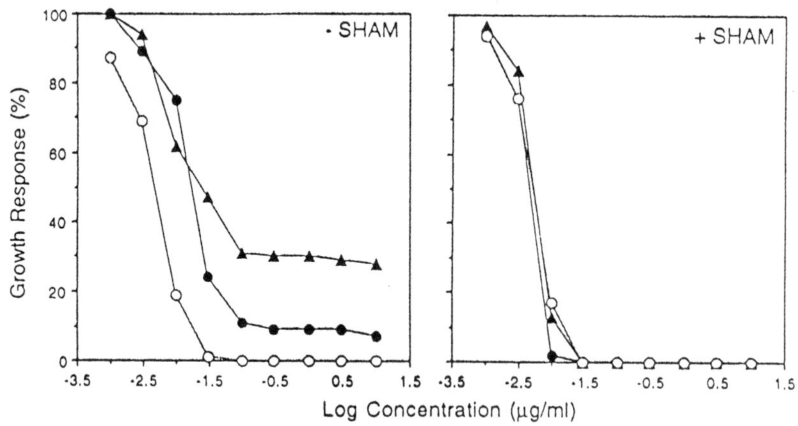

Figure 4. Response of Magnaporthe grisea *mycelium of a wild-type strain (triangles) and two REMI mutants (closed and open circles) to azoxystrobin in the absence and presence of SHAM at a concentration of (100 µg/ml).*

Table I. Sensitivity of *Venturia inaequalis* to kresoxim-methyl[a]

	ED_{50} (µg/ml)		
Isolate	*Conidia*	*Mycelium*	*Scab Control*
S-56-88	1.0	0.3	0.1
KNH-6-95	0.8	0.005	0.2

[a] Data are from Olaya et al. (*19*).

The apparent discrepancy between *in vitro* and *in vivo* responses of pathogens to strobilurin fungicides were explained by a model proposed for *M. grisea* (*17*). According to this model, plant secondary products such as flavone act as scavengers of active oxygen species. They will lower the intracellular concentration of active oxygen species required for the induction of alternative respiration and, thus, will prevent activation of this mechanism of in a host-pathogen environment. Supporting data have been provided for *V. inaequalis* (*19, 24*) and *B. cinerea* (*20*), and preliminary results indicated that rice, as the host of *M. grisea*, contained products synergizing the *in vitro* effects of metominostrobin (*27*).

Although the chemical nature and the ubiquitous presence of plant products antagonizing the induction of alternative respiration have not been determined in any detail, this model implies that such plant-derived antioxidants must penetrate into fungal cells before they can antagonize the induction of alternative respiration. It can, thus, be expected that restricted uptake of such natural antagonists would allow respective phenotypes to freely induce alternative respiration even under *in vivo* conditions of disease control. Mechanisms of restricted uptake of compounds into fungal cells mediated by the family of ABC transporters have been identified, and their involvement in conferring resistance to the sterol demethylation inhibitor class of fungicides has been described (*28*). Similar transport systems might also restrict the uptake of plant antioxidants allowing respective phenotypes to resist strobilurin-related fungicides *via* induction of alternative respiration. Experimental evidence for the existence of such phenotypes is lacking at the present time.

Mechanism of Strobilurin Resistance: Target Site Mutations

Mutations of Cytochrome *b* Conferring Strobilurin Resistance

Mutations of the cytochrome *b* target site have long been recognized as a potential mechanism of fungal resistance to strobilurin-related inhibitors. A total of 13 such target site mutants have been identified for *S. cerevisiae*, a fungus lacking the alternative oxidase gene. Respective mutations were localized within two interhelical loops of cytochrome *b*, both involved in the architecture of the Q_P center (*29, 30*). Consequences of mutational amino acid exchanges were largely variable levels of mitochondrial resistance to strobilurin-related center Q_P inhibitors and variable effects on the functional integrity of the cytochrome bc_1 complex (Table II). Fungal mitochondria naturally resistant to strobilurin-related inhibitors were also isolated from the two strobilurin-producing basidiomycetes *Mycena galopoda* and *Strobilurus tenacellus* (*31*).

Resistance based on a mutational change of the target site in ascomycetes have, thus far, only been found for a laboratory mutant of the pathogen *V. inaequalis* (*32*). Conidia were derived from an isolate highly sensitive during the stage of conidia germination (Table I), and germinating conidia were

Table II. Impact of mutational amino acid exchanges in cytochrome *b* of *Saccharomyces cerevisiae* on resistance to strobilurin-related inhibitors and on the efficiency of electron transfer[a]

Inhibitor	F129L	G137E	N256Y	N256Y	N256K	N256I	L275S[b]
			Resistance Factor[c]				
Myxothiazol	105[d]	13[d]	23	3[d]	1	2	35 [d]
Strobilurin A	5	4	21	10	1	2	-[e]
Oudemansin	3	2	-	11	-	-	-
MOA-Stilbene	2	2	68	5	3	5	-
Azoxystrobin	2	68	-	1	-	-	2
KM[f]	9	180	-	1	-	-	24
Famoxadone	13	4	-	0.3	-	-	2
			Relative V_{max}[g]				
	46 [d]	150	11	26	44	63	71

[a]Data are from Jordan et al. (*6*), Geier et al. (*38, 41*), Giessler et al. (*42*) and Thierbach and Michaelis (*39*).
[b]Mutational amino acid exchanges.
[c]ED_{50} values of mitochondrial respiration in comparison to wild-type.
[d]Means of several figures.
[e]Data not reported.
[f]Kresoxim-methyl.
[g]V_{max} of electron flux through complex III in comparison to wild-type (=100).

selected on a high dose of kresoxim-methyl subsequent to UV mutagenesis. Under these conditions, mutants resisting high doses of kresoxim-methyl were altered in their capacity of expressing alternative respiration during the stage of conidia germination, but their cytochrome *b* target site remained unchanged (*32*). Benomyl-resistant mutants based on mutations of the *β*-tubulin target site (*33*) could be generated with ease under the same conditions of mutagenesis, suggesting that the procedures employed were suitable for selecting mutants with altered fungicide-binding sites. However, a strobilurin-resistant mutant with an altered cytochrome *b* target site could be selected by including the alternative oxidase inhibitor SHAM in the selection medium. This result is best explained by the mitochondrial origin of the gene encoding the cytohrome *b* target site. Mutations can only be expected to affect one of the multiple genes present in the population of mitochondria. Induction of alternative oxidase as a response to strobilurin action by the majority of sensitive mitochondria would lead to the circumvention of the target site in both sensitive and resistant mitochondria (Figure 2). In response, respiration would become identical in both sensitive and resistant mitochondria, and the process of replacing sensitive by resistant mitochondria would be halted.

The mutant of *V. inaequalis* selected in the presence of kresoxim-methyl and SHAM was highly resistant to the mixture and was distinguished by a glycine to alanine exchange in position 143, a position located in one of the interhelical loops also affected by mutations in *S. cerevisiae* (Figure 5). An identical G143A exchange leading to mitochondria resisting strobilurin-related inhibitors had been reported for mouse cell cultures (*34*). Likewise, the cytochrome *b* sequences of the strobilurin producer *Mycena galopoda* (*31*) and the sea urchin *Paracentrotus lividus* (*35*), both naturally resistant to this class of inhibitors, were also distinguished by an alanine in the same position (Figure 5).

An alanine was also conserved in chloroplast cytochrome b_6 known to be almost insensitive to strobilurin-related inhibitors (*36*) in spite of sequence similarities shared with mitochondrial cytchrome *b* (Figure 5). These multiple lines of evidence indicate that the G143A exchange identified for *V. inaequalis* is likely to confer resistance to this class of respiration inhibitors. Interestingly, a G143A exchange has not been identified for resistant mutants of *S. cerevisiae* (Figure 5), for mutants of the green alga *Chlamydomonas reinhardtii* (*45*) and the naturally resistant strobilurin producer *Strobilurus tenacellus* (*31*). Explanations for this apparent discrepancy between mutation sites have not yet been provided.

Resistance Risks Inherent to Target Site Mutations

The example of *V. inaequalis* indicates that strobilurin resistance mediated by a mutational change of the cytochrome *b* target is feasible, but also that a single resistant mitochondrion must first be selected to dominance within a single cell before a resistant phenotype will emerge. Because alternative oxidase would be incorporated into both sensitive and resistant mitochondria, this selection process will be slow or prevented, if the cytochrome *b* target site is circumvented by the rapid induction of alternative respiration (Figure 2). The

```
A

121                   143              160
s FILMIVTAFLGYVLPYGQMSLWGATVITNLMSAIPWIGQD V.inaequalis
r                       A
s LFAVMATAFMGYVLPWGQMSFWGATVITNLLSAIPYIGTT Mouse
r                       A
r LVLMMAIGFLGYVLPFGQMSLWAATVITNLLSSIPVFGQD M.galopoda
r FLVTVLTAFVGYVLVWGRMSFWAATVIANLVTAVPCVGTT Sea Urchin

s FTLTIATAFLGYCCVYGQMSHWGATVITNLFSAIPFVGND S.cerevisiae
r         L
r             Y
r                 R
r                 E
r                 V
r                      K
r                      T


B
              143
s LGYVLPYGQMSLWGATVITNLMSAIPWIGQDIVEFLWGGFSVN V.inaequalis
r              A
r TGYSLPWDQIGYWAVKIVTGVPDAIPVIGSPLVELLRGSASVG Rice
r TGYSLPWDQIGYWAVKIVTGVPDAIPVIGSPLVELLRGSASVG Spinach
r TGYSLPWDQIGYWAVKIVTGVPDAIPVIGSPLVELLRGSASVG Arabidopsis
r TGYSLPWDQVGYWAVKIVTGVPDAIPGVGGFIVELLRGGVGVG C. reinhardtii
```

Figure 5. Comparison of cytochrome b *amino acid sequences of* Venturia inaequalis *sensitive (s) or resistant (r) to kresoxim-methyl with corresponding sequences of organism mutated or naturally resistant to strobilurin-related inhibitors. A. Mitochondrial cytochrome b.* B. *Chloroplast cytochrome b_6. Data are assembled from published accounts (29-45). NCBI GenBank accession numbers are AAB95255 (*Venturia inqequalis*), P00158 (mouse), 2206325B (*Mycena galopoda*), NP 008133 (sea urchin), X84042 (*Saccharomyces cerevisiae*), P12123, (rice), P00165, (spinach), BAA94415 (*Arabidopsis thaliana*) and Q00471 (*Clamydomonas reinhadtii*).*

multiple lines of evidence described above have indicated, however, that for the majority of fungal baseline isolates, the induction of alternative respiration is 'silenced' under *in vivo* conditions of disease control. Consequently, the replacement of sensitive mitochondria by resistant ones could take place during the control of diseases with strobilurin fungicides.

The speed of intracellular mitochondrial selection will, most likely, be determined by the relative 'fitness' of resistant mitochondria. Most of the distinct mutations of cytochrome *b* conferring strobilurin resistance to *S. cerevisiae* were accompanied by reduced electron flow through the cytochrome bc_1 complex (Table II). Such functionally impaired mitochondria might be selected during exposure to strobilurin-related inhibitors, but their frequencies are likely to decline as soon as cells are relieved from inhibitor action. Indications that mitochondrial populations return to a sensitive stage exist for the G143A mutation characterized for *V. inaequalis* (*32*). Heteroplasmic stages of organelle DNA with impact on the level of resistance conferred by cytochrome *b* mutations have also been reported for mouse cell cultures (*34*) and the chloroplast-encoded target site of triazine herbicides (*46*). However, homoplasmic stages of organelle DNA conferring resistance can be reached. For example, the transformation of plant chloroplasts stably expressing foreign genes mandates the selection of respective chloroplast DNA to a homoplasic stage (*47*). Once such homoplasmic stages of 'strobilurin-resistant' cytochrome *b* DNA are reached, respective phenotypes would stably express the trait of strobilurin resistance. If fitness penalties inherent to a mutated target site were absent or not substantial, such phenotypes would then persist in populations, and they would be selected during the control of diseases with strobilurin fungicides.

Outlook

At present, two mechanisms of strobilurin resistance are likely to be of relevance in the development of practical resistance to strobilurin fungicides: the induction of alternative respiration and the mutational change of the cytochrome *b* target site. The assessment of risks inherent to alternative respiration remains speculative. The induction of this mechanism is well documented under *in vitro* conditions, but it is not necessarily active during the control of diseases. However, mechanisms allowing fungal phenotypes to even resist strobilurin fungicides through alternative respiration under *in vivo* conditions are likely to exist. For example, the uptake of plant antioxidants antagonizing the induction of alternative respiration might be actively lowered, or the concentration of active oxygen species known to act as inducers of alternative oxidase might be decreased by the overexpression of fungal superoxide dismutase or catalase. Although experimental evidence is lacking at present, it can be expected that such multigenic mechanisms of resistance will individually confer a relatively small level of resistance, and that respective phenotypes will remain to be controlled at high inhibitor doses. This phenomenon has been demonstrated for fungicides inhibiting sterol demethylation, and a high dose strategy was sucessful in slowing the selection of phenotypes responding fully resistant at a low dose of these fungicides (*48*).

The second mechanism of resistance expected to be relevant will be a mutational change of the cytochrome *b* target site, which has been reported for several organisms. As described above, a target site mutation could also be generated in the plant pathogen *V. inaequalis* (Figure 5), but only after the 'equalizing' effect of alternative respiration was eliminated in order to mimic the situation found under conditions of disease control (*32*). The mitochondrial origin of the cytochrome *b* gene implies that resistant mitochondria will first have to be selected within the cells of individual phenotypes before such phenotypes will acquire full resistance. The speed of this selection process will be determined by both the level of resistance conferred by a particular mutation and the functional integrity of the mutated cytochrome *b* target site. The data available for the most exhaustively investigated fungus *S. cerevisiae* imply that both parameters are largely variable and dependent on both the nature of the inhibitor and the mutation site (Table II).

Preventive anti-resistance strategies coping with target site mutations will, thus, be affected by all three parameters. Although relatively small levels of resistance similar to the scenario described for *S. cerevisiae* (Table II) could be counteracted by a high dose strategy, this strategy would fail for large factors of resistance reported for some of the mutation and inhibitor combinations identified (Table II). The G143A mutation reported for *V. inaequalis* was found to confer such a high level of resistance (*32*), the reoccurrence of sensitive mitochondria in response to the relief of *V. inaequalis* cells (*32*) from the inhibitor also indicated a respiratory penalty inherent to the mutation. Such functional penalties could be utilized in preventive anti-resistance measures. The selection of mutated during disease cycles controlled with strobilurin-related fungicides could be reversed by the counter-selection of more 'fit' sensitive mitochondria during periods of disease control by other means, and a relatively stable situation could be reached. This scenario would call for a restricted number of strobilurin treatments per season, an anti-resistance strategy currently recommended by FRAC. At present, a high dose strategy and the restriction of treatments per season appear to constitute advisable anti-resistance measures.

Acknowledgements.

The work with *V. inaequalis* was supported, in part, by BASF AG, Ludwigshafen, Germany and by the United States Department of Agriculture

References

1. Von Jagow, G.; Link, T. A. *Methods Enzymol.* **1986,** *126*, 253-271.
2. Anke, T. *Can. J. Bot.* **1995,** *73,* S940-S945.
3. Bautement, K.; Clough, J. M.; de Fraine, P. J.; Godfrey, R. A. *Pestic. Sci.* **1991,** *31*, 499-519.

4. Baldwin, B. C.; Clough, J. M.; Godfrey, C. R. A.; Godwin J. R.; Wiggins, T. E. In *Modern Fungicides and Antifungal Compounds*; Lyr, H.; Russel, P. E.; Sisler, H. D., Eds.; Intercept, Andover, 1996; pp 69-77.
5. Ypema, H. L.; Gold, R. E. *Plant Dis.* **1999,** *83*, 4-19.
6. Jordan, D. B.; Livingston, R. S.; Bisaha, J. J.; Duncan, K. E.; Pember, S. O.; Picollelli, M. A.; Schwartz, R. S.; Sternberg, J. A.; Xiao-Song, T. *Pestic. Sci.* **1999,** *55*, 105-118.
7. Clough, J. M.; Anthony, V. M.; de Fraine, P. J.; Frasser, T. E. M.; Godfrey, A. R. A.; Godwin, J. R.; Youle, D. In *Eighth International Congress of Pesticide Chemistry: Options 2000*; Ragsdale, N. N.; Kearny, P. C.; Plimmer, J. R., Eds.; American Chemical Society, Washington DC, 1995, pp 59-73.
8. Sauter, H.; Ammermann, E.; Benoit, R.; Brand, S.; Gold, R. E.; Grammenos, W.; Köhle, H.; Lorenz, G.; Müller, B.; Röhl, F.; Schirmer, U.; Speakman, J. B.; Wenderoth, B.; Wingert, H. In *Antifungal Agents – Discovery and Mode of Action*; Dixon, G. K.; Copping, L. G.; Hollomon, D. W., Eds.; BIOS Scientific Publishers, Oxford, 1995, pp 173-192.
9. Zheng, D.; Köller, W. *Curr. Genet.* **1997,** *32*, 361-366.
10. Day, D. A.; Whelan, J. ; Millar A. H.; Siedow, J. N.; Wiskich, J. T. *Aust. J. Plant Physiol.* **1995,** *22*, 497-509.
11. Vanlerberghe G. C.; McIntosh, L. *Annu. Rev. Plant Physiol. Molec. Biol.* **1997,** *48*, 703-734.
12. Mitzutani, A.; Miki, N.; Nanba, K. *Pestic. Biochem. Physiol.* **1998,** *60*, 187-194.
13. Maxwell, D.-P.; Wang, Y. and McIntosh, L. *Proc. Natl. Acad. Sci. USA* **1999,** *96*, 8271-8276.
14. Joseph-Horne, T.; Hollomon, D. W. *Pest Manag. Sci.* **2000,** *56*, 24-30.
15. Yukioka, H.; Inagaki, S.; Tanaka, R.; Katoh, K.; Miki, N.; Mitzutani, A.; Masuko, M. *Biochim. Biophys. Acta* **1998,** *1442,* 161-169.
16. Huh, W. K.; Kang, S. O. *J. Bacteriol.* **1999,** *181*, 4098-4102.
17. Mitzutani, A.; Miki, N.; Yukioka, H.; Tamura, H.; Masuko, M. *Phytopathology* **1996,** 295-300.
18. Ziogas, B. N.; Baldwin, B. C.; Young, J. E. *Pestic. Sci.* **1997,** *50*, 28-34.
19. Olaya, G.; Zheng, D.; Köller, W. *Pestic. Sci.* **1998,** *54,* 230-236.
20. Tamura, H.; Mizutani, A.; Yukioka, H.; Miki, N.; Ohba, K.; Masuko, M. *Pestic. Sci.* **1999,** *55,* 681-686.
21. Yukioka, H.; Tanaka, R.; Inagaki, S.; Katoh, K.; Miki, N.; Mizutani, A.; Masuko, M.; Kunoh, H. *Fungal Genet. Biol.* **1997,** *22*, 221-228.
22. Yukioka, H.; Inagaki, S.; Tanaka, R.; Katoh, K.; Miki, N.; Mizutani, A.; Masuko, M. *Ann. Phytopathol. Soc. Jpn.* **1998,** 282-286.
23. Shi, Z.; Christian, D.; Leung, H. *Phytopathology* **1995,** *85*, 329-333.
24. Olaya, G.; Köller, W. *Plant Disease* **1999,** *83*, 274-278.
25. Olaya, G.; Köller, W. *Pestic. Sci.* **1999,** *55*, 1083-1088.
26. Li, Q.; Ritzel, R. G.; McLean, L. L. T.; McIntosh, L.; Ko, T.; Bertrand, H.; Nargang, F. E. *Genetics* **1996,** *142*, 129-140.
27. Kume, R.; Nikawa, M.; Kawaguchi, A.; Yasumi, K.; Masuko; M.; Shiraishi, T. *Ann. Phytopathol. Soc. Jpn.* **1997,** 110-112.
28. De Waard. *Pestic. Sci.* **1997,** *51*, 271-275.

29. Colson, A.-M. *J. Bioenerg. Biomem.* **1993,** *25*, 211-220.
30. Di Rago, J. P.; Mermann-Le Denmat, S.; Paques, F.; Risler, J.; Netter, P.; Slonimski, P. P. *J. Molec. Biol.* **1995,** *248*, 804-811.
31. Kraiczy, P.; Haase, U.; Gencic, S.; Findt, S.; Anke, T.; Brandt, U.; von Jagow, G. *Eur. J. Biochem.* **1996,** *235,* 54-63.
32. Zheng, D.; Olaya, G.; Köller, W. *Curr. Gent.* **2000,** in press.
33. Koenraadt, H.; Somerville, S. C.; Jones, A. L. *Phytopathology* **1992,** *82,* 1348-1354.
34. Howell, N.; Gilbert, K. *J. Molec. Biol.* **1988,** *203*, 607-618.
35. Degli Eposti, M.; de Vries, S.; Crimi, M.; Ghelli, A.; Patarnello, T.; Meyer, A. *Biochim. Biophys. Acta* **1993,** *1143*, 243-271.
36. Rich, P. R.; Madgwick, S. A.; Wiggins, T. E. *Biochem. Soc. Transact.* **1994,** *22*, 217-221.
37. Di Rago, J. P.; Colson, A.-M. *J. Biol. Chem.* **1988,** *263*, 12564-12570.
38. Geier, B. M.; Schägger, H.; Brandt, U.; Colson, A.-M.; von Jagow, G. *Eur. J. Biochem.* **1992,** *208,* 375-380.
39. Thierbach, G.; Michaelis, G. *Mol. Gen. Genet.* **1982,** *186,* 501-506.
40. Lemesle-Meunier, D.; Brivet-Cheviolette, P. di Rago, J. P.; Slominski, P.; Bruel, C.; Tron, T.; Forget, N. *J. Biol. Chem.* **1993,** *268*, 15626-15632.
41. Geier, B. M.; Haase, U.; von Jagow, G. *Biochem. Soc. Transact.* **1994,** *22*, 203-209.
42. Giessler, B. M.; Schägger, H.; Brandt, U.; Colson, A.-M.; von Jagow, G. *Eur. J. Biochem.* **1994,** *222*, 147-154.
43. Bruel, C.; Manon, S.; Guérin, M.; Lemesle-Meunier, D. *J. Bioernerg. Biomembr.* **1995,** *27,* 527-539.
44. Bruel, C.; di Rago, J. P.; Slominski, P. P.; Lemesle-Meunier, D. *J. Biol. Chem.* **1995,** *270*, 22321-22328.
45. Bennoun, P.; Delosme, M.; Godehardt, I.; Kück, U. *Molec. Gen. Genet.* **1992,** *234*, 147-154.
46. Frey, J. F. *Nature* **1999,** *398*, 115-116.
47. Staub, J. M.; Garcia, B.; Graves, J.; Hajdukiewicz, P. T. J.; Hunter, P.; Nehra, N.; Paradkar, V.; Schlitter, M.; Carroll, J. A.; Spatola, L.; Ward, D.; Ye, G.; Russell, D. A. *Nature Biotechnol.* **2000,** *18,* 333-338.
48. Köller, W.; Wilcox, W. F. *Plant Dis.* **1999,** *83*, 857-863.

Chapter 15

Case Studies of Fungicide-Resistance Management in Japan

Tetsuo Amano and Toshiyuki Takeda

Pesticide Research Section, ZEN-NOH Agricultural Research and Development Center 5–5–1 Higashiyahata, Hiratsuka, Kanagawa 254–0016 Japan

Developing ment of fungicide-resistance is one of the serious factors of fungicide efficacy and may damage crop-yield and -quality. The pesticide research section of ZEN-NOH Agricultural R&D Center, an experiment and research institute for pesticide of Japan Agricultural Cooperative Associations, has been conducting experiments for ranking fungicides and evaluating their characteristics and others, taking much consideration of farmers as fungicide users. The experimental results from our study on fungicide-resistant are summarized as follows :"DMIs(DeMethylation-Inhibitors) decreased sensitivity to *Erysiphe graminis* by spraying the same fungicide"; "DMIs sensitivity baselines and sensitivity distributions of isolates sampled over Japan were estimated in strawberry-powdery-mildew fungi"; and "Strobirulin-resistant fungi occurred in cucumber powdery and downy mildew "

© 2002 American Chemical Society

I. Sensitivity of *Erysiphe graminis* to DMIs

In Hokkaido where the most wheat is cultivated in Japan, we studied how the triadimefon spraying will affect the sensitivity of *Erysiphe graminis* to triadimefon.

Materials and methods

The study was conducted in Hokkaido, the northern district in Japan, from 1991 through 1994. The wheat powdery mildew is one of the principal blights in wheat cultivation in Hokkaido. Since the variety "CHIHOKU" in particular, cultivated principally when we conducted this study, was very susceptible to this blight, its control is critical.

In Hokkaido, people sow wheat seeds at the end of September and harvest next summer. Triadimefon is sprayed in November just before snowfall or in early summer of May and June to protect wheat either from snow blight or powdery mildew. Two experimental fields were set; one was sprayed with triadimefon twice or thrice, the other without triadimefon. The experimental fields have 500 square meters each and they were apart by 120 m.

The sensitivity assay was conducted using leaf-segment method(Fig.1). After wheat had been grown under powdery-mildew-free conditions, the central part of the 2nd leaf was cut into 10-15 mm to be put side by side on a wet filter paper in a petri dish onto which the powdery mildew conidia, disinfected and air-dried, were tapped down to inoculate. Such leaves were floated on the surface of given concentrations of the chemical to examine how the blight macula was developed after exposure by 7000 Lux-12 hours for 6 days at 20℃. Those results were used to determine MIC and EC_{50} by regression analysis (1).

Results and conclusions

The results were shown in Fig2. The figure2 has dates as abscissa and EC_{50} for *E. graminis* sensitivity to triadimefon as vertical axis. The arrow shows the timing of triadimefon spraying.

In untreated-field, the sensitivity of *E.graminis* to triadimefon did not change during the experimental period. On the other hand, in a triadimefon-treated field, the sensitivity of *E.grrminis* to triadimefon decreased with the pesticide spray in growing season; after the harvest, however, the sensitivity recovered partly, but not perfectly.Such results suggested, in the actual field, the triadimefon sensitivity gradually decreased by continual application of triadimefon. In view of this result and others, the local government in Hokkaido provided their instructions to spray DMIs only once a year.Thus, DMIs can avoid the tremendous decrease of their efficacy (2,3) .

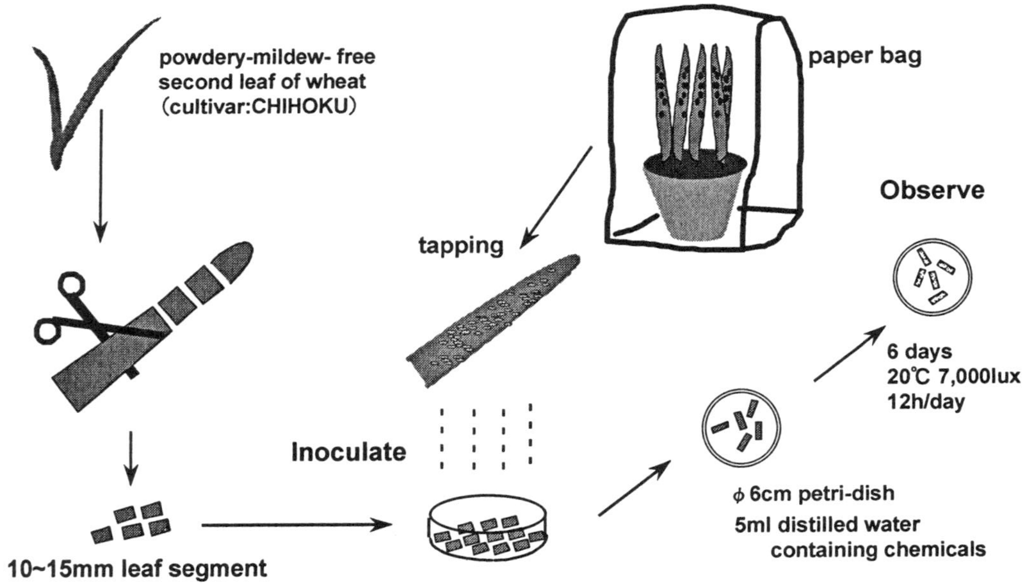

Fig.1 Method for evaluating sensitivity of *E. graminis* f. sp. *tritici* against systemic fungicides

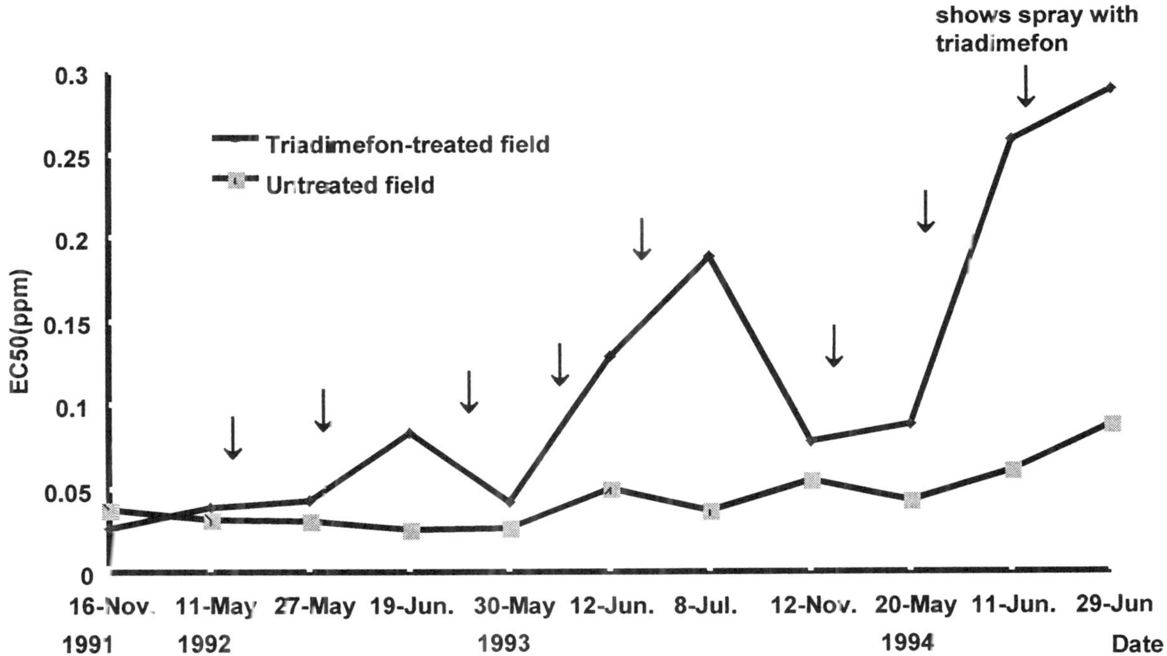

Fig.2 Change of sensitivity of *E. graminis* f. sp. *tritici* to triadimefon in an untreated field and triadimefon-treated field

II Sensitivity of Strawberry Powdery-mildew fungi to DMIs

As strawberry powdery mildew fungi infect on fruits as well as leaves, they cause a great damage for quality and quantity of strawberry and the major cultivars of strawberry in Japan are all highly susceptible to powdery mildew, so it is necessary to control them by fungicides.

Materials and methods

The strawberry-powdery-mildew sensitivity baseline to DMIs has not been studied in Japan, and in 1994, when we started sampling strawberry-powdery-mildew fungi DMIs had already been used widely as a leading agent for controlling the strawberry-powdery-mildew. Accordingly, it was necessary to deduce its sensitivity baseline to DMIs from the data obtained in the area far from field for strawberry cultivation. We first sampled a powdery-mildew fungus from the mountain. It is a parasite on *Fragaria nipponica* belonging to the same genus *Fragaria* as the cultivar strawberry. Then we examined how this fungus were susceptible to DMIs, or if it could be determined as the same fungus as the pathogen of strawberry-powdery-mildew. There could be indicated its parasiticalness to all of 6 cultivar strawberries(Table 1). In addition, the cultivar-strawberry-powdery-mildew fungus inoculation to *Fragaria nipponica* showed some pathopoiesis. In addition, the morphological microscope investigation revealed no difference between the cultivar-strawberry-powdery-mildew fungi and that parasiting *Fragaria nipponica.* Thus, the powdery-mildew fungus sampled from *Fragaria nipponica* might be considered as one of the strawberry-powdery-mildew fungi.

The pesticide sensitivity assay was carried out using leaf-disk method (Fig.3).

A leaf disk was cut out with 9 mm-diameter cork borer from a compound leaf from the strawberry sapling grown under powdery-mildew-fungus-free conditions. The leaf disks were put side by side on a wet filter paper in the 13 cm-diameter petri dish, and the powdery-mildew fungal conidia were brushed off on the leaf disks to be inoculated. After 7-10 days since the inoculated leaves had been floated on the surface of a DMIs solutions of given concentrations. The development of the blight macula was examined. The DMIs sensitivities were studied for 31 strawberry-powdery-mildew fungal isolates sampled from various areas in Japan from 1994 through 1998. As DMIs, 3 agents of triflumizole, fenarimol and microbutanil, which are all widely used in strawberry cultivation, were served for the study(4).

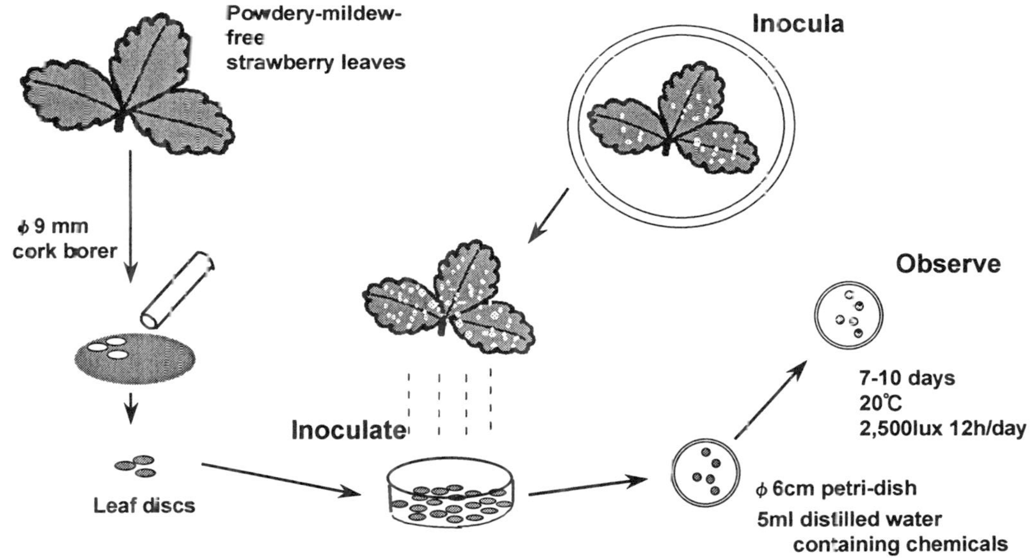

Fig.3 Method for evaluating sensitivity of *S. ahanis* against systemic fungicides

Results and conclusions

EC_{50}s for each DMI agent showed a one-peak frequency distribution; trifrumizol, fenarimol and microbutanil were applicable for 0.2-5.6, 0.2-7.7 and 0.02-2.0 ppm, respectively (Figs.4).

The arrows shown in Fig.4 indicates EC_{50} of the powdery-mildew fungus parasiting *Fragaria nipponica* described earlier. This isolate showed the highest sensitivity to DMIs. Consequently, when it was employed as a sensitivity baseline to DMIs for computing R-factor, triflumizol, fenarimol and microbutanil have, respectively, 113, 86 and 99 as R-factor for the least sensitive isolate. It has been reported that DMIs will permit less sensitive mutants to readily develop, and, infact, there aer decreased sensitivity and controlling effect in the strawberry-powdery-mildew fungi. As to the cucumber-powdery-mildew fungi that are decreasing in their sensitivities further than the strawberry-powdery-mildew fungi, it has been confirmed that such isolates have their R-factors not less than 100 in terms of sensitivity baseline and allow the controlling effects of DMIs to decrease.

Positive correlations EC_{50} of 3 DMIs tested for the strawberry-powdery-mildew fungi, suggesting that these sensitivities cross over each other (Figs.5).

In addition, when we conducted the experiment to examine the controlling effects of the DMIs in green-houses contaminated with such fungal strains having low sensitivities to DMIs, effects of DMIs were low and those of other fungicides were high (Table2). Consequently, the DMI sensitivity is decreasing for the strawberry-powdery-mildew fungus.

III Sensitivity of Cucumber Powdery and Downy mildew fungi to strobilurins

The strobirulin fungicides, having both a wide anti-fungal spectra and high effects, are very promising as new antifungal agents. In Japan, kresoxim-methyl and azoxystrobin were registered in December 1997 and in April 1998, respectively, from which dates each fungicidal has been used by farmers. However, cucumber powdery and downy mildew allowed the fungi having remarkably low sensitivities to them to be detected from the isolates sampled, respectively, in March 1999 and in June 1999. (5,6,7)

Materials and methods

As in strawberry-powdery-mildew fungi described above, the fungicidal sensitivity assay is conducted using leaf-disk method in which on the surface of a

Table 1. Pathogenicity of the isolates from *F. nipponica* to 6 strawberry caltivars

Host Plant / Test Plants	Disease severity					
	Toyonoka	Nyohoh	Reikoh	Himiko	Harunoka	Fukuba
Fragaria nipponica	3.8	4.6	4.8	3.2	2.8	5.0

*:Disease severity was evaluated using leaf discs of strawberry by scoring with a scale of 0 to 5.

Table 2. Efficacy of some fungicides against strawberry powdery mildew (field test)

Fungicides	Fungicide group	Dilution ratio	% of diseased berry	Protective value
Kresoxim-methyl FL(47%)	strobilurins	x5000	0.0	100.0
Amistar(azoxystrobin)FL(20%)	strobilurins	x2000	7.0	71.3
MepanipyrimFL(40%)	anilinopyrimidines	x3000	2.8	88.6
MyclobutanilWP(10%)	DMI	x4000	7.5	69.2
TriflumizoleWP(30%)	DMI	x3000	11.5	52.5
Untreated	-	-	24.2	-

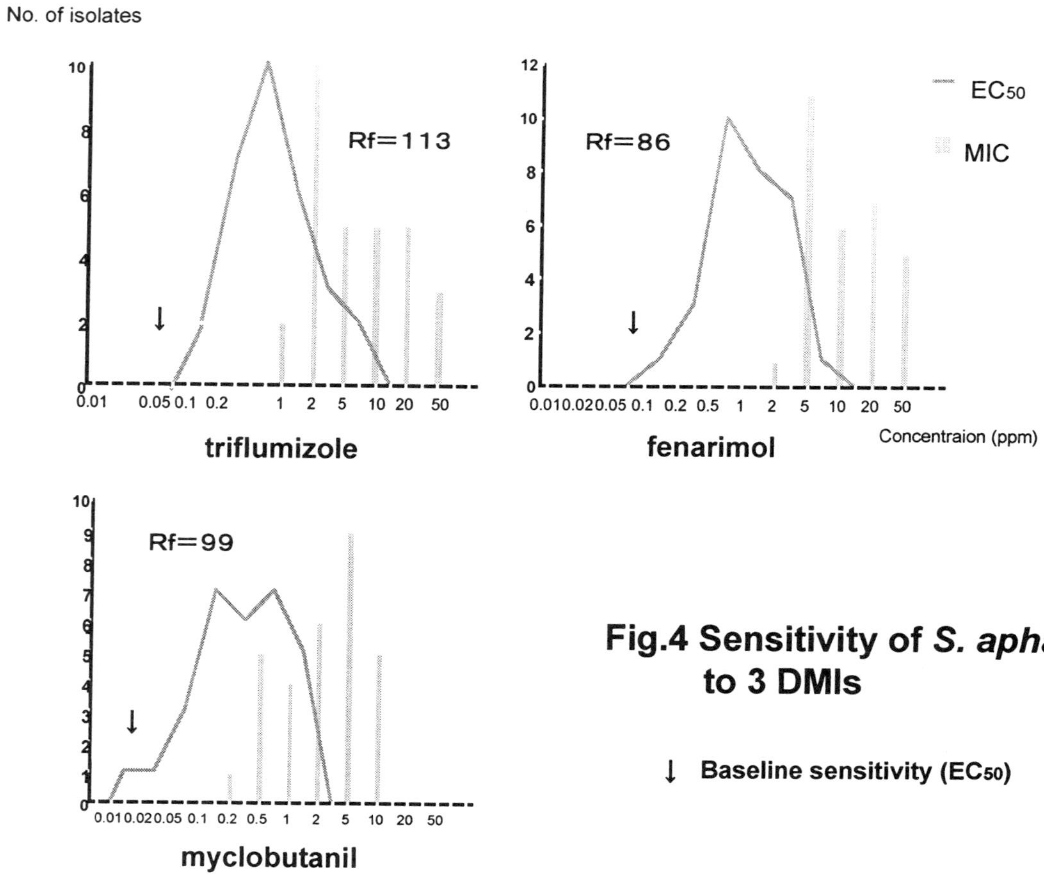

Fig.4 Sensitivity of *S. aphanis* to 3 DMIs

↓ **Baseline sensitivity (EC_{50})**

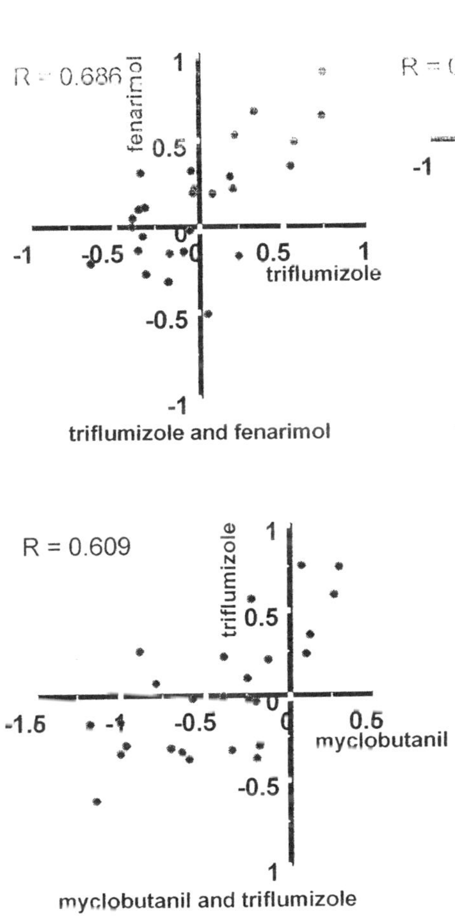

Fig.5 EC_{50} correlations of 3 DMIs(EC50)

Table 3. Sensitivity of *Sphaerotheca fuliginea* to strobilurins

Isolates	azoxystrobin		kresoxim-methyl
	MIC(ppm)	EC_{50}(ppm)	MIC(ppm)
I-1	10	3.6	>20
I-2	10	3.5	>20
T-1	10	4.6	>20
T-2	10	4.1	>20
S-1	20	5.6	>20
K-7	0.01	0.0042	0.5

・K-7 is the sensitive standard isolate of ZEN-NOH

Table4. Sensitivity of *Pseudoperonospora cubensis* isolate to strobilurins

Isolates	azoxystrobin		kresoxim-methyl	
	MIC(ppm)	EC_{50}(ppm)	MIC(ppm)	EC_{50}(ppm)
MA	>30	-	>30	-
FS	0.3	0.069	30	7

MA is *P. cubensis* isolate from Miyazaki agricultural experimental station to strobilurins

FS is sensitive standard isolate of ZEN-NOH

given-concentration-fungicidal diluent were floated leaf disks cut down from such leaves cultivated under powdery-mildew-free conditions.

Results and conclusions

The sensitivities of tested cucumber-powdery-mildew and -downy mildew fungi to strobilurins were found out to be still lower than those of the control fungi(Table3,4).

In an experiment employing pot saplings, no controlling effect was observed over such fungi by strobilurins at the usual serving concentrations. Thus, it was concluded that the fungi tested here are strobilurins-resistant.

In future as an experimental and research institute for Agricultural Cooperative Associations in Japan, we will make every possible effort to carry out expcriments and researches to protect crops from thc damage by fungicide-resistance.

References

1. Nakazawa,Y. , *Annals of the Phytopathological Society of Japan*, 91,58, 12-13

2. Takeuchi,T. , *Annals of the Phytopathological Society of Japan*, 1992,58,148

3. Takeuchi,T.;Shimizu,M;Nakazawa,Y;Kodama,F., *HOKUNOU*, 1993,vol.60,2,174-178

4. Nakazawa,Y.;Uchida,K.;Inoue,J.;Takeda,T.;Amano,T.;Kita,N., *ICPP98* Paper Number 2.2.6

5. Fujita,N.;Hamada,K.;Inagaki,M.;Enoyoshi,T.;Miyahara,T.;Kojiguchi,S.;Ogasawara,K;Kadota,G.;Dale,S.M.;Takamatsu,S., Annals of the Phytopathological Society of Japan, 1999,65,692

6. Ishii,H.;Noguchi,K.;Tomita,Y.;Umemoto,S.;Nishimura,K., *Annals of the Phytopathological Society of Japan*, 1999,65,655

7. Takeda,T.;Kawagoe,Y.;Uchida,K.;Fuji,M.;Amano,T., *Annals of the Phytopathological Society of Japan*, 1999,65,655

Chapter 16

DNA-Based Approaches for Diagnosis of Fungicide Resistance

Hideo Ishii

National Institute for Agro-Environmental Sciences, Tsukuba, Ibaraki 305–8604, Japan

The increasing use of modern diagnostic tools to detect and quantify diseases and fungicide resistance will play an important part in improving the rational use of fungicides within integrated disease management systems. When testing fungicide resistance, bioassays including mycelial growth and spore germination tests *in vitro*, and inoculation tests on plants are usually employed. However, such methods are generally time-consuming, laborious, and allow relatively few fungal isolates to be monitored. Furthermore, biotrophs such as powdery mildews, downy mildews or slow-growing fungi like *Venturia* spp. are difficult to handle in general. Development of novel methods will therefore be required to quickly test fungicide resistance. In this article, the author will summarize the DNA-based techniques for diagnosis of fungicide resistance, which have already appeared in literatures. The suggested molecular mechanism of resistance to the newly developed strobilurin fungicides is also mentioned.

 © 2002 American Chemical Society

Identification of Fungal Species which differ in Fungicide Sensitivity

Within the past several years, new developments in technology have enabled the molecular diagnosis of biotic plant diseases (1). In laboratories, RFLP (Restriction Fragment-Length Polymorphism), Southern hybridization, PCR (Polymerase Chain Reaction), and other techniques are now commonly used for this purpose. Some agrochemical companies are also interested in the development and practical application of new methods for disease diagnosis on crops. Using these methods, the diagnostic laboratories, consultants or farmers themselves will be able to know the presence of pathogens on crops earlier facilitating optimization of countermeasures. Such approaches are particularly important when aiming to minimize the input of pesticides into the environment.

Molecular Diagnosis of *Colletotrichum acutatum*

It is possible that individual fungal species belonging to the same genus differ in their sensitivities to a certain fungicide. As such an example, two anthracnose fungi, *Colletotrichum gloeosporioides* and *C. acutatum*, each known to be the cause of grapevine ripe rot (2, 3), are cited here. *C. gloeosporioides* is inherently sensitive to benzimidazoles and insensitive to diethofencarb. But benzimidazole-resistant isolates of this species have also been found in fields where this group of fungicides was applied and negative cross-resistance between benzimidazoles and diethofencarb was observed. Another species *C. acutatum* is inherently insensitive to both benzimidazoles and diethofencarb (4). Occasionally these two species are distributed concomitantly in vineyards.

Conidia representative of *C. gloeosporioides* are cylindrical with an obtuse apex. Those typical *C. acutatum* are fusiform and each end of the conidium is abruptly tapered. However, even in the same isolate, considerable variation was recognized in the size and shape of conidia. In these pathogens, the species-specific primers annealing to ITS regions of ribosomal DNA had already been reported (5, 6). *C. acutatum* was successfully diagnosed using the species-specific primer not only from isolated pure culture of the fungus but also from infected young grapevine tissue (2). Partial sequences of the fungicide targeted β-tubulin genes of *C. acutatum* and *C. gloeosporioides* were compared. One-base difference at codon 198 (AAG to GAG) was found in an isolate of *C. acutatum* when compared to benzimidazole-sensitive isolate of *C. gloeosporioides* (Figure 1). The same difference was also found between

	Codon 198 200
Grapevine ripe rot	
C. acutatum	
G1-1 5 (R, R)	TCG GAC AAG ACC TTC TGC
G5-1 7 (R, R)	TCG GAC AAG ACC TTC TGC
C. gloeosporioides	
GC-2 (S, R)	TCG GAC GAG ACC TTC TGC
Strawberry anthracnose	
C. acutatum	
Na91-016 (R, R)	TCG GAC AAG ACC TTC TGC
C. gloeosporioides	
NM-1 (S, R)	TCG GAC GAG ACC TTC TGC

Figure 1. Partial alignments (codons 196 to 201) of the β-tubulin gene sequence from isolates of Colletotrichum acutatum and C. gloeosporioides which differed in their sensitivities to benzimidazoles and diethofencarb. Sites of base change are underlined.

isolates of *C. acutatum* and *C. gloeosporioides* collected from strawberry. PCR primer 'HR.RIII' (5'-ACCAGCTTGTCGAGAATTCGGACAA-3') was designed to specifically detect the allele in isolates of *C. acutatum* encoding insensitivity to both benzimidazoles and diethofencarb. Using this primer in combination with the reverse primer 'D' (5'-GCTGGATCCTATTCTTTGGGTCGAACAT-3') from Koenraadt (7), positive bands of β-tubulin gene fragments were produced from both infected grapevine tissues and DNA purified from fungal culture.

It is likely that the application of benzimidazole fungicides may select and increase populations of *C. acutatum* in the field. In fact, *C. acutatum* was frequently detected in an experimental vineyard in which benomyl had been sprayed for the control of ripe rot during a dormant period (8). Furthermore, control efficacy of benomyl treatment dropped (9). Monitoring of fungicide insensitivity will be necessary as a mixture of diethofencarb with thiophanate-methyl, which is not effective against *C. acutatum*, is now used in vineyards for the control of other pathogens like gray mold.

Mechanisms of Fungicide Resistance and Molecular Diagnosis

Basic studies on the molecular genetics on fungicide resistance are generally indispensable for the application of DNA-based techniques to monitor resistance in an agricultural environment.

Benzimidazole Resistance

Molecular techniques have been used in the work on benzimidazole resistance in *Botrytis*, *Venturia* spp. etc. since the mechanism of resistance was well characterized (10). RFLP, Southern hybridization, PCR, and SSCP (Single-Strand Conformation Polymorphism) analysis have been applied.

Japanese pear scab, caused by *Venturia nashicola*, is an important disease as scab-resistant pear cultivars are not yet commercially available. Accordingly, growers tend to spray fungicides many times a year. However, mycelial growth of this fungus is extremely slow, and it usually takes several months to get fungicide sensitivity data after the fungus is isolated from diseased plant materials. The response of *V. nashicola* to benzimidazole fungicides is classified into four groups, *i.e.* sensitivity, weak resistance, intermediate resistance, and high resistance. Classical genetic studies showed that allelic mutations govern the resistance and different levels of resistance are due to different combinations of these mutations at the single chromosomal locus (11).

Most field resistant isolates of plant pathogenic fungi tested, exhibited codon changes either at positions 198 or 200 in β-tubulin genes (Table 1). This was also the case for *V. nashicola* (12). Substitution of glutamic acid (GAG) at codon 198 by alanine (GCG) resulted in high resistance to the benzimidazole fungicide carbendazim. This mutation seemed to be further involved in negative cross-resistance to diethofencarb although biochemical evidence couldn't be obtained from this fungus. In an isolate showing intermediate resistance to carbendazim, the amino acid at codon 200 was altered from phenylalanine (TTC) to tyrosine (TAC). The single-base substitution in the targeted β-tubulin gene will change the physicochemical nature of the binding site for benzimidazoles. Actually, decreased binding of radiolabelled carbendazim to a β-tubulin fraction from resistant isolates has been demonstrated (13).

Using the sequence data on the β-tubulin gene of *V. nashicola*, restriction sites were identified (12). The base change (GAG to GCG) in codon 198 created a *Tha*I restriction site together with a *Hga*I site. Two restriction fragments were generated from the PCR amplified DNA samples bearing such a mutation after treatment with each enzyme. The resistant isolate, which showed negative cross-resistance to diethofencarb, was clearly identified by this PCR-RFLP method (Figure 2). Allele-specific PCR (ASPCR) primers were also designed based on the sequence difference and used for the diagnosis of resistance. Mutations in codon 198 or codon 200 were both identified by ASPCR. Figure 3 shows detection of the allele encoding high benzimidazole resistance with increased sensitivity to diethofencarb.

With *V. nashicola* DNA, SSCP was first introduced in the diagnosis of fungicide resistance (12). When denatured, single strand DNA having one base-change can be separated on a gel according to the difference of electrophoresis patterns. We separated PCR products of the β-tubulin gene fragment using capillary gel electrophoresis and fluorescence detection system. A mutation at either codon 198 or codon 200 was identified by the difference in migration time of the DNA sample. When the mixture of DNA samples from both highly carbendazim-resistant and –sensitive isolates was applied onto a gel, each sample was separated as a result of the different conformation of each DNA strand (Figure 4).

DMI Resistance

The primary mode of action of sterol demethylation-inhibiting fungicides (DMIs) is thought to be as follows: binding to the haem of the sterol

14 α-demethylase, inhibition of C14 demethylation of eburicol, accumulation of precursor sterols and reduction in ergosterol and/or other desmethyl sterols and eventually, the disruption of membrane function.

A number of DMIs are registered for the control of Japanese pear scab. First generation DMIs such as triflumizole, bitertanol, and fenarimol have been used for about fifteen years with others like difenoconazole, hexaconazole, and imibenconazole also currently used. Monitoring of fenarimol sensitivity revealed a shift to lower sensitivity in some isolates of *V. nashicola* collected from DMI-treated pear orchards (14, 15). Fortunately, fenarimol showed adequate control against less fenarimol- sensitive isolates in inoculation tests on pear seedlings, indicating that the performance of this fungicide was still maintained in the field. However, the long time necessary for testing fungicide resistance interferes with quick and accurate decision-making required to minimize fungicide control failure. Therefore, to develop molecular diagnosis of resistance, fragments of the sterol 14 α-demethylase gene were amplified by PCR directly using DNA isolated from scab-infected pear materials. However, no differences in deduced amino acid sequences of the sterol 14 α-demethylase were found in isolates which differed in their sensitivities to DMIs (16). Mechanisms other than point mutation in the targeted gene will probably be involved in the resistance.

Several mechanisms have been proposed for DMI resistance in phytopathogenic fungi. First, point mutation in the sterol 14 α-demethylase gene has been reported in laboratory mutants of citrus blue mold *Penicillium italicum* (17) and field isolates of grapevine powdery mildew (18). In the latter fungus *Uncinula necator*, resistance factors (RF) to the DMI triadimenol ranged from 1.8 to 26.0. A single mutation, leading to the substitution of a phenylalanine residue for a tyrosine residue at position 136, was always associated with resistance higher than RF >10. In contrast, no mutation was found in weakly resistant (RF <5) isolates. Tyrosine has a hydroxyl group that does not exist in phenylalanine. The loss of the hydroxyl group in the active site of eburicol 14 α-demethylase should increase the hydrophobicity of the site without affecting its configuration. Increased hydrophobicity at the active site might reduce the affinity of the enzyme for its inhibitor (a polar molecule with a hydrophobic skeleton) (18).

ASPCRs yielded an amplified DNA fragment of the expected size from all highly resistant isolates of *U. necator* used (18). However, no amplification occurred with DNAs from some moderately resistant isolates (10>RF>5). As the authors wrote, it is possible that some resistant isolates possess another point mutation(s) in the eburicol 14 α-demethylase gene that may result in different levels of DMI resistance. Alternatively, accumulation of different mechanisms may result in an increase in the level of resistance.

Table 1 Point mutation and deduced amino acid substitutions in the β-tubulin gene for field isolates of phytopathogenic fungi with resistance to benzimidazole fungicides

Fungal species	Phenotype* (B/D)	Amino acids in position		Reference
		198	200	
Botrytis cinerea	S/HR	Glu	Phe	39
	HR/S	Ala	Phe	39
	HR/WR	Val	Phe	40
	HR/HR	Lys	Phe	41
	MR/HR	Glu	Tyr	41
Helminthosporium solani	S/	Glu	Phe	42
	I/	Gln		42
	R/	Gln		42
	R/	Ala		42
Penicillium digitatum	S/HR	Glu	Phe	7
	IR/HR	Glu	Tyr	43
	HR/HR	Lys	Phe	7
Rhynchosporium secalis	S/HR	Glu	Phe	44
	HR/S	Gly	Phe	44

Tapesia yallundae	S/HR	Glu	Phe	45
	HR/S	Ala	Phe	45
	HR/S	Gly	Phe	45
	HR/HR	Lys	Phe	45
	IR/HR	Glu	Tyr	45
	(LR/HR)**	Glu	Phe	45
	LR/HR	Gln	Phe	45
Venturia inaequalis	S/HR	Glu	Phe	7
	LR/HR	Glu	Phe	7
	MR/S	Gly	Phe	7
	MR/HR	Glu	Tyr	7
	HR/HR	Lys	Phe	7
	VHR/S	Ala	Phe	7
V. nashicola	S/HR	Glu	Phe	12
	WR/HR	Glu	Phe	12
	IR/HR	Glu	Tyr	12
	HR/S	Ala	Phe	12
	HR/HR	Lys	Phe	12

*Sensitivity to benzimidazoles (B) or / diethofencarb (D): S, sensitive; I, intermediate; R, resistant; WR, weakly resistant; LR, lowly resistant; IR, intermediately resistant; MR, moderately resistant; HR, highly resistant; VHR, very highly resistant.

**Isolates in this phenotype class had a change at codon 240, substituting a leucine with a phenylalanine.

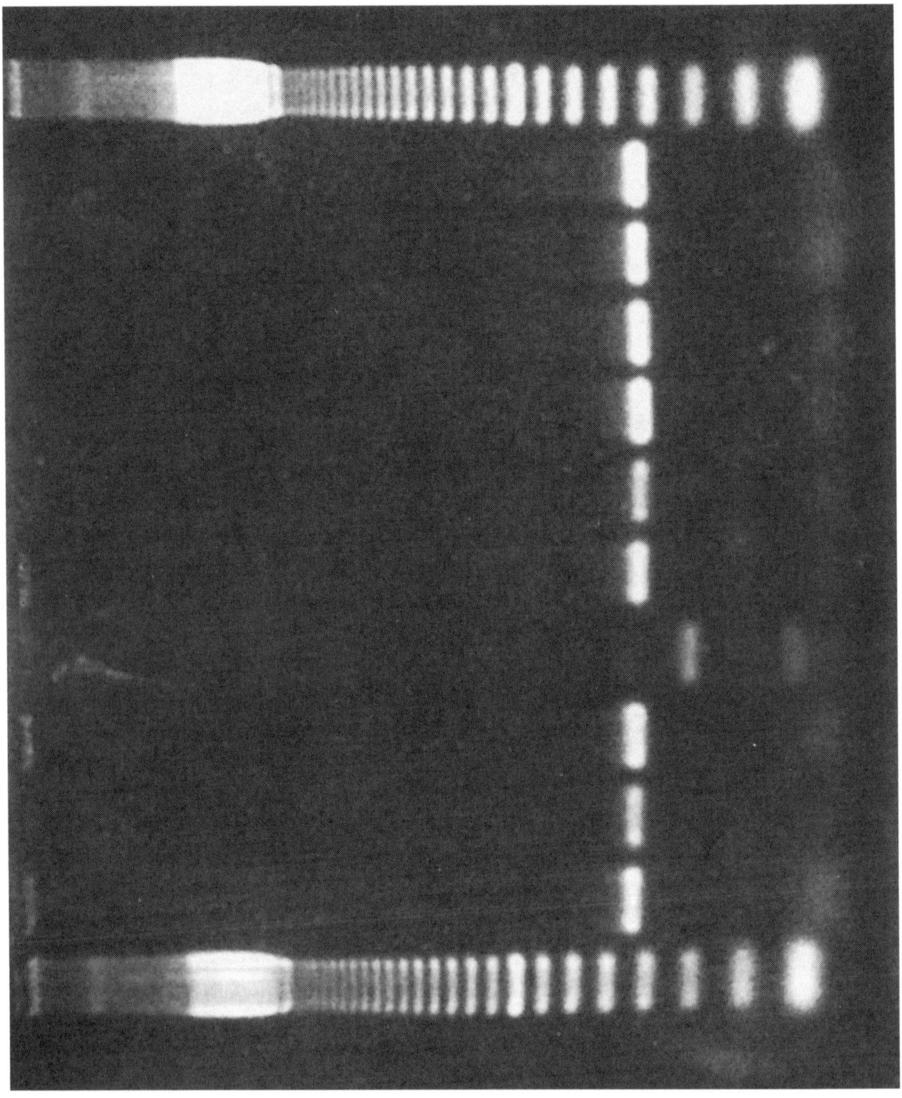

Figure 2. Diagnosis of high benzimidazole resistance associated with negative cross-resistance to dietofencarb in Venturia nashicola, the cause of Japanese pear scab, using PCR-RFLP. The β-tubulin gene fragment was PCR-amplified from fungal genomic DNA purified from pure culture. The products were treated with restriction enzyme ThaI, separated by agarose gel electrophoresis, and stained with ethidium bromide. The enzyme recognized one base change from GAG to GCG at codon 198 occurring in isolates, highly benzimidazole-resistant and diethofencarb-sensitive, but not in other isolates, and generated two digestion fragments after treatment. Tracks are M: 100bp ladder, 1 to 5: ThaI digests of the PCR products from an isolate showing 1; S/HR, 2; WR/HR, 3; IR/HR, 4; HR/S, 5; HR/HR and 6 to 10: enzyme-untreated products from isolates showing 6; S/HR, 7; WR/HR, 8; IR/HR, 9; HR/S, 10; HR/HR (See legend of Table 1).

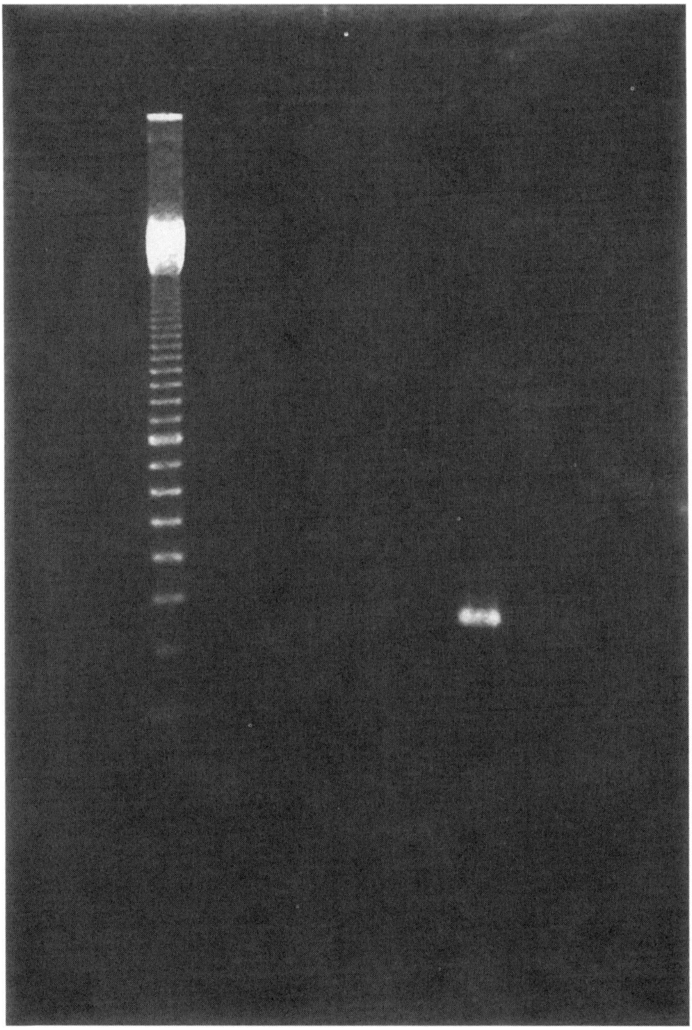

Figure 3. Diagnosis of high benzimidazole resistance associated with negative cross-resistance to dietofencarb in Venturia nashicola using allele-specific PCR (ASPCR). The β-tubulin gene fragment was first amplified by PCR from fungal genomic DNA and nested PCR was further carried out. A primer was designed and used to specifically amplify A to C mutation at codon 198 occurring in isolates highly benzimidazole-resistant and diethofencarb-sensitive, but not in other isolates. Tracks are M: 100bp ladder, 1 to 5: amplifications from isolate showing 1; S/HR, 2; WR/HR, 3; IR/HR, 4; HR/S, 5; HR/HR (See legend of Table 1).

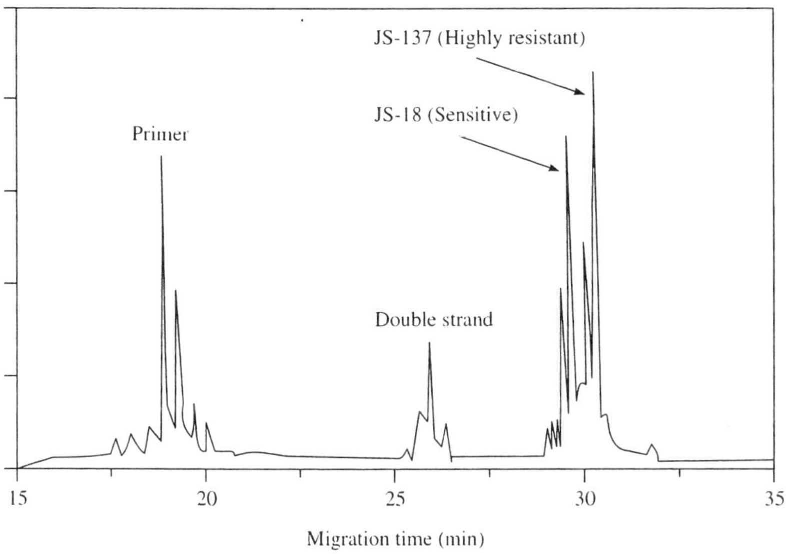

Figure 4. Detection of one base change in the β-tubulin gene of Venturia nashicola by single-strand conformation polymorphism (SSCP) analysis. PCR products of the β-tubulin gene amplified from benzimidazole-sensitive and highly benzimidazole-resistant isolates were separated by capillary gel elctrophoresis under denaturing conditions, and the migration time of each allele measured. The allele (GCG_{198}) encoding high benzimidazole resistance could be distinguished from the wild type sensitive allele (GAG_{198}) by differences in migration time of DNA.

In barley powdery mildew (*Blumeria graminis* f. sp. *hordei* = *Erysiphe graminis* f. sp. *hordei*), two alleles for the eburicol 14 α-demethylase gene were identified (19). DMI-sensitive reference isolates exhibited a TAT (tyrosine) codon at position 136 whereas the isolate collected from a DMI-treated field showed a TTT (phenylalanine) codon at that position. ASPCR experiments confirmed that other isolates obtained from DMI-treated fields carried the same mutation in the gene. Furthermore, only minute amounts of fungal material (1 mg dry weight of conidia) were used for the PCR. It is possible that, as found in *P. italicum* and *U. necator*, substitution of a phenylalanine residue for a tyrosine residue may cause resistance to DMIs in *B. graminis* f. sp. *hordei*.

Another mechanism of DMI resistance, increased efflux of fungicides from fungal cells, is also thought to be a key factor, and overproduction of the multidrug transport proteins, *i.e.* P-glycoproteins or ABC transporters may be involved (20, 21). Furthermore, overexpression of the sterol 14 α-demethylase and the cytochrome P450 reductase genes have also been proposed as mechanisms of DMI resistance (22). Recently, tandem repeat of a transcriptional enhancer was found in the promoter region, upstream of the sterol 14 α-demethylase gene in DMI-resistant isolates of *P. digitatum* (23). Constitutive expression of the gene was about 100-fold higher in resistant isolates than in sensitive isolates and resulted in higher DMI resistance. Also in *V. inaequalis*, higher expression of the sterol 14 α-demethylase gene, observed in DMI-resistant field isolates, was correlated with the presence of an insertion located upstream of the gene, suggesting that overexpression of the target-site gene is an important mechanism of resistance (24). However, other mechanisms of resistance also appeared to exist as other isolates of this fungus showed resistance in the absence of the insertion.

Strobilurin Resistance

The strobilurin fungicides such as azoxystrobin and kresoxim-methyl have been developed and are now widely used in practice in many countries as they have broad controlling spectrum against many fungal diseases on various crops. These fungicides bind to the subunit protein of the cytochrome bc_1 complex of the electron transport system in the mitochondrial respiratory chain thereby causing inhibition of fungal respiration (25).

The first report of field resistance to this class of fungicides came from wheat powdery mildew (*B. graminis* f. sp. *tritici* = *E. graminis* f.sp. *tritici*) in northern Germany in 1998 (26, 27). Subsequently, the resistance problem has also occurred in powdery mildew (*Podoshaera fusca* = *Sphaerotheca fuliginea*) and downy mildew (*Pseudoperonospora cubensis*) on cucumber in

Japan (28, 29, 30) and Taiwan. Between late 1997 and early 1998, cucumber growers started to use azoxystrobin and kresoxim-methyl for controlling these diseases. Most of them followed the manufacturers' recommendation and applied their products only a couple of times per crop in alternation with other fungicides, which possess different modes of action. Later on, however, growers noticed rapid decline of fungicide efficacy against powdery mildew.

Bioassay data clearly showed that the emergence of strobilurin-resistant isolates was causing the decline in fungicide efficacy. Resistant isolates were approximately 100 to 1,000 times less sensitive to azoxystrobin than sensitive isolates in leaf disc tests. Currently, these isolates seem to be widely distributed in cucumber growing areas where control failure of the fungicides has been reported.

Subsequently, strobilurin-resistant isolates have also been found in cucumber downy mildew. In inoculation tests, neither azoxystrobin nor kresoxim-methyl controlled resistant isolates despite each fungicide being applied at the recommended concentration. Cross-resistance between these two fungicides was demonstrated.

Fragments of the mitochondrial cytochrome *b* gene were PCR-amplified from fungal DNA and their sequences analyzed to understand the molecular mechanism of resistance. A single point mutation, *i.e.* one base change at codon 143, was found in the gene from resistant isolates of *P. cubensis* and *P. fusca* (31, 32). Substitution of glycine at codon 143 in cytochrome *b* by alanine seemed to result in high resistance to strobilurins. The same mutation has also been found in resistant isolates of *Mycosphaerella fijiensis*, the cause of black Sigatoka of banana (33) and wheat powdery mildew (34). These data strongly suggested that resistance to strobilurin fungicides in field isolates of plant pathogens due to a target site alteration.

To develop rapid testing methods for strobilurin sensitivity of the obligate parasite *P. cubenis*, one disc (1 cm in diameter) was cut from a leaf bearing zoosporangia and used for DNA extraction with a mortar and pestle. After purifying the pathogen DNA by a commercially available kit, a fragment of the cytochrome *b* gene was successfully amplified by PCR. Subsequently, products were treated with the restriction enzyme *Ita*I, which should specifically recognize the mutated sequence in the fragment. It was shown that this PCR-RFLP method could be used in practice for rapid monitoring of pathogen isolates resistant to strobilurin fungicides (Figure 5, 35).

*Ita*I

Resistant isolates: ↓

CCTTGGGGACAAATGAGTTTTTGGG<u>C</u>TGCAACTGTTATTACTAA

Sensitive isolates:

CCTTGGGGACAAATGAGTTTTTGGGGTGCAACTGTTATTACTAA

Figure 5. Partial nucleotide sequences of the mitochondrial cytochrome b gene from strobilurin-resistant and -sensitive isolates of Pseudoperonospora cubensis. Nucleotide sequences encoding residue 143 of the cytochrome b protein are highlighted by shaded boxes. The site of point mutation is underlined. Arrow indicates the restriction site of the enzyme ItaI, which recognises the resistant genotype.

Future Prospect

Since the author wrote the chapter on "Monitoring of fungicide resistance in fungi: biological to biotechnological approaches" earlier, molecular and biochemical techniques have advanced rapidly (36). These include nonradioactive oligonucleotide probes instead of radio-labelled ones for Southern hybridization, PCR primers for RAPD (random amplified DNA polymorphism) analysis (37), and development of real-time PCR equipment etc.

The Home-Grown Cereals Authority (HGCA), of the UK, has provided funding for projects on diagnostics for crop disease assessment. Novartis (now Syngenta) is developing a range of crop disease diagnostics based on PCR technology. One of these is a new test for eyespot in cereals and it identifies the different eyespot strains, W-type (*Tapesia yallundae*) and R-type (*T. acuformis*). Diagnosis before the symptoms appear in particular, will enable fungicide dose and timing to be matched to disease levels. Diagnostic test kits will allow farmers to justify their disease control measures and make them cost-effectively.

The portable advanced nucleic acid analyzer (ANAA) has been described recently (38). This instrument could overcome a number of limitations *e.g.* size, weight, speed and others, which commercial spectrofluorometric thermal cyclers possess for on-the-go testing of samples. For field use, real-time PCR is the most practical as it doesn't require time-consuming post-PCR manipulation such as gel electrophoresis. The experimental results demonstrated that the total time required to detect a bacterial sample could be as little as 7 minutes.

Information on the molecular mechanisms of fungicide resistance will also be included in the diagnostic test kits for pathogens so that early, accurate identification of pathogens and quantification of isolates resistant to fungicides would be effectively achieved in future.

References

1. Schots, A.; Dewey, F. M.; Oliver, R. *Modern Assays for Plant Pathogenic Fungi: Identification, Detection and Quantification*, 1994, CAB International, Wallingford, 1-267.
2. Ishii, H.; Iwamoto, S.; Nishimura, K.; Fukaya, M. *Proc. 1998 Brighton Conf. - Pests & Diseases*, 1998, 529-534.

3. Yamamoto, J.; Sato, T.; Tomioka, K. *Ann. Phytopathol. Soc. Jpn.*, 1999, *65*, 83-86.
4. Sato, T.; Uematsu, S.; Iijima, A.; Koganezawa, H. *Nippon Kingakukai Kaiho*, 1998, *39*, 35-44.
5. Mills, P. R.; Sreenivasaprasad, S.; Brown, A. E. *FEMS Microbiol. Lett.*, 1992, *98*, 137-144.
6. Sreenivasaprasad, S., Sharada, K.; Brown, A. E.; Mills, P. R. *Plant Pathol.*, 1996, *45*, 650-655.
7. 7. Koenraadt, H.; Somerville, S. C.; Jones, A. L. *Phytopathology*, 1992, *82*, 1348- 1354.
8. 8. Fukaya, M.; Ishii, H.; Takahashi, I. *Ann. Phytopathol. Soc. Jpn.*, 1998, *64*, 394- 395.
9. Fukaya, M. *Abstr. 11th Symp. Res. Commit. Fungicide Resistance (Phytopathol. Soc. Jpn.)*, 2001, 9-15.
10. Davidse, L. C.; Ishii, H. *Modern Selective Fungicides* (H. Lyr ed.), 1995, Gustav Fischer, Jana, 305-322.
11. Ishii, H.; Yanase, H.; Dekker, J. *Med. Fac. Landbouww. Rijksuniv. Gent*, 1984, *49/2a*, 163-172.
12. Ishii, H.; Kamahori, M.; Hollomon, D. W.; Narusaka, Y. *Abstr. Resistance '97*, 1997.
13. Ishii, H.; Davidse, L. C. *Proc. 1986 Br. Crop Protection Conf. - Pests & Diseases*, 1986, 567-573.
14. Ishii, H. *Pestic. Sci.*, 1997, *51*, 383-386.
15. Tomita, Y.; Ishii, H. *Pestic. Sci.*, 1998, *54*, 150-156.
16. Cools, H. J.; Ishii, H.; Butters, J. A.; Hollomon, D. W. *J. Phytopathol.*, submitted.
17. De Waard, M. A. *ACS Symp. Ser.*, 1996, *645*, 62-71.
18. Délye, C.; Laigret, F.; Corio-Costet, M-F. *Appl. Environ. Microbiol.*, 1997, *63*, 2966-2970.
19. Délye, C.; Bousset, L.; Corio-Costet, M-F. *Curr. Genet.*, 1998, *34*, 399-403.
20. De Waard, M. A. *Pestic. Sci.*, 1997, *51*, 271-275.
21. Nakaune, R.; Adachi, K.; Nawata, O.; Tomiyama, M.; Akutsu, K.; Hibi, T. *Appl. Environ. Microbiol.*, 1998, *64*, 3983-3988.
22. Van den Brink, H. J. M.; Van Nistelrooy, H. J. G. M.; De Waard, M. A.; Van den Hondel, C. A. M. J. J.; Van Gorcom, R. F. M. *J. Biotechnol.*, 1996, *49*, 13-18.
23. Hamamoto, H.; Hasegawa, K.; Nakaune, R.; Lee, Y. J.; Makizumi, Y.; Akutsu, K.; Hibi, T. *Appl. Environ. Microbiol.*, 2000, *66*, 3421-3426.
24. Schnabel, G.; Jones, A. L. *Phytopathology*, 2001, *91*, 102-110.
25. Clough, J. M.; Godfrey, C. R. A. *Fungicidal Activity* (D. Hutson & J. Miyamoto eds.), 1998, John Wiley & Sons, Chichester, 109-148.

26. Anonymous. *Agrow*, 1998, *318*, 11.
27. Erichsen, E. *Getreide*, 1999, *5*, 44-46.
28. Ogasawara, K.; Enoyoshi, T.; Miyahara, T.; Kojiguchi, S.; Kadota, G.; Takamatsu, S.; Dale, S. M. *Ann. Phytopathol. Soc. Jpn.*, 1999, *65*, 655.
29. Takeda, T.; Kawagoe, Y.; Uchida, K.; Fuji, M.; Amano, T. *Ann. Phytopathol. Soc. Jpn.*, 1999, *65*, 655.
30. Ishii, H.; Noguchi, K.; Tomita, Y.; Umemoto, S.; Nishimura, K. *Ann. Phytopathol. Soc. Jpn.*, 1999, *65*, 655.
31. Ishii, H. *Abstr. 10th Symp. Res. Commit. Fungicide Resistance (Phytopathol. Soc. Jpn.)*, 2000, 43-51.
32. Heaney, S. P.; Hall, A. A.; Davies, S. A.; Olaya, G. *Proc. BCPC Conf. - Pests & Diseases*, 2000, 755-762.
33. Sierotzki, H.; Parisi, S.; Steinfeld, U.; Tenzer, I.; Poirey, S.; Gisi, U. *Pest Manag. Sci.*, 2000, *56*, 833-841.
34. Sierotzki, H.; Wullschleger, J.; Gisi, U. *Pestic. Biochem. Physiol.* 2000, *68*, 107-112.
35. Ishii, H.; Sugiyama, T.; Nishimura, K.; Ishikawa, Y. *Modern Selective Fungicides* (H.-W. Dehne *et al.* eds.), Intercept, Andover, *in press.*
36. Ishii, H.; *Molecular Methods in Plant Pathology* (R. P. Singh & U. S. Singh eds.), 1995, CRC Press, Boca Raton, 493-503.
37. Délye, C.; Laigret, F.; Corio-Costet, M-F. *Pestic Sci.*, 1997, 51, 309-314.
38. Belgrader., P.; Benett, W.; Hadley, D.; Richards, J.; Stratton, P.; Mariella Jr., R.; Milanovich, F. *Science*, 1999, *284*, 449-450.
39. Martin, L. A.; Fox, R. T. V.; Baldwin, B. C.; Connerton, I. F. *Proc. 1992 Brighton Crop Protection Conf. - Pests & Diseases*, 1992, 207-214.
40. Yamada, M.; Ishii, H.; Nakazawa, Y. *Ann. Phytopathol. Soc. Jpn.*, 1997, *63*, 225.
41. Yarden, O.; Katan, T. *Phytopathology*, 1993, *83*, 1478-1483.
42. McKay, G. J.; Cooke, L. R. *FEMS Microbiol. Lett.*, 1997, *152*, 371-378.
43. Adachi, K.; Namba, S.; Tsuchizaki, T.; Hibi, T. *Ann. Phytopathol. Soc. Jpn.*, 1994, *60*, 744.
44. Wheeler, I. E.; Kendall, S. J.; Butters, J.; Hollomon, D. W. *Pestic. Sci.*, 1995, *43*, 201-209
45. Albertini, C.; Gredt, M.; Leroux, P. *Pestic. Biochem. Physiol.*, 1999, *64*, 17-31.

Chapter 17

Novel Method for Monitoring Fungicide Resistance in *Botrytis cinerea*

Kiyotsugu Okada and Shin-ichi Kusakari

Osaka Prefectural Agricultural and Forestry Research Center, Habikino-city, Osaka 583–0862, Japan

Botrytis cinerea causes serious damage on many crops. Selection and timely application of fungicides are most important for the control of gray mold. To monitor the distribution of resistant strains in the field, the following selective medium for *B. cinerea* isolates was developed. Composition of this medium is $MgSO_4.7H_2O$, 0.5 g ; $NaNO_3$, 2g ; K_2HPO , 1g ; L-glutamine, 2g ; fructose, 10g ; and agar, 15g in 1000 ml of DW for basal medium. Streptomycin sulfate, 300mg ; triphenyltin acetate, 1.5mg ; copper oxide chloride, 25mg ; Rose Bengal, 25mg were added as inhibitors. The air-borne spores of *B. cinerea* were easily trapped on this medium by using a novel air sampler. Further, adding the fungicides to this selective medium could monitor more easily and rapidly the fungicide resistance of *B. cinerea.* This method was more convenient for the diagnosis of resistant strains as compared to a conventional mycelial growth method.

Introduction

The gray mold disease caused by *Botrytis cinerea* is an important disease, which induces serious damage to a variety of crops. Low temperature and high humidity in greenhouses promote the occurrence of gray mold disease during winter and spring in Japan. In vegetables such as eggplant and cucumber, control of this pathogen is difficult due to fungicide resistance. Fungicide resistance develops because of the continuous use of certain fungicides. Although

 © 2002 American Chemical Society

fungicides with different action mechanisms are applied in rotation, isolates which show the resistance to two or more fungicides are commonly found.

It is necessary to determine the distribution of the fungicide resistance of *B. cinerea* to control gray mold disease. To examine fungicide resistance of *B. cinerea*, a special technique is necessary. The process of sample collecting, isolating and culturing of *B. cinerea*, and examining the results requires two to three weeks. On sites such as an agricultural extension center or a plant protection office, a more efficient method of promptly and easily obtaining results is needed. Therefore, we developed the spore trap method for monitoring fungicide resistance. This method consists of a selective medium and the spore collection machine.

Materials and methods

The selective medium for *B. cinerea*

The composition of the selective medium (SBc) for *B. cinerea* is shown in Table 1. The autoclaved basic medium was cooled down to 50 ., and the inhibitors were added to it. The composition of the basic medium is $MgSO_4 \cdot 7H_2O$, 0.5g; $NaNO_3$, 2g; K_2HPO_4, 1g; L-glutamine, 2g; fructose, 10g; and agar, 15g in 1000 ml of distilled water. The composition of the inhibitors is streptomycin sulfate, 300mg; triphenyltin acetate, 1.5mg; copper oxide chloride, 25mg; and Rose Bengal, 25mg. Triphenyltin acetate 150mg is dissolved in 100ml of dimethyl sulfoxide (DMSO). Adjusting the pH of the selective medium is not necessary (pH is 5.8 ± 0.2).

In SBc and already reported selective media (Lorbeer amd Tichelaar, 1970; Kritzman and Netzer, 1978; Bardinell et al., 1989; Kerssies, 1990), the isolation efficiency of *B.cinerea* was compared. A spore suspension of *B. cinerea* was spread on each selective medium tested in this experiment. In order to collect the air-borne spore samples of gray mold in greenhouses, on the selective media uniformly and efficiently, we developed a spore collection machine (Fig.1). The spore samples were collected in the eggplant and cucumber greenhouses using this machine. It was set at half of plant height of object crops, and was moved vertically for 3-5 minutes between 8:00-11:00 a.m. With this device, the spore-containing air was sucked through a perforated surface and passed over the selective medium. The spores were trapped on the medium, and the air was exhausted with a motor fan through holes set on partition board behind the medium. The power of this device was supplied by a 12-volt battery. The tests were performed at least twice with the media selective for the spores.

Use of the spore trap method for monitoring fungicide resistance

Fungicide resistance was monitored by the spore trap method and the SBc medium with fungicides. The colony formation rates of each fungicide resistant

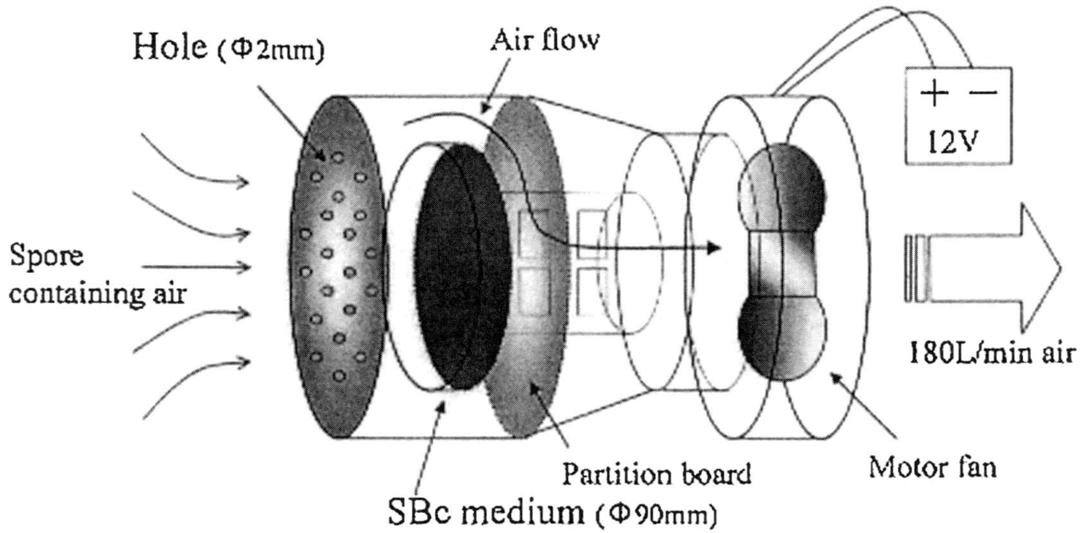

Figure 1. The model of the spore collection machine

phenotype were examined on SBc medium containing fungicides such as 10ppm benomyl, 5ppm procymidone or 10ppm diethofencarb. A spore suspension (100 CFU/ml) was spread on each medium, and the number of colonies which appeared at 20. after culturing for 5 days was counted. Next, two petri dishes were divided into three sections as shown in Figure 2, and the SBc medium containing fungicides was poured into the separate divisions. The spore trap method was applied as described earlier. The proportions (%) of the fungicide resistant strains were calculated from the number of colonies of *B. cinerea* formed on the SBc medium with fungicide(s) divided by that of colonies which grew on SBc medium without fungicide.

Result and Discussion

Comparison of the selective media for *B. cinerea*

After 4 days culturing on SBc medium at 20-25. under dark conditions, colonies of *B. cinerea* were observed by the naked eye. The colony surface of *B. cinerea* was downy and magenta or grayish white, and the back of colony was also magenta. The shapes of the colonies were uniform circles, 0.5-2.0mm diameter when cultured for 7 days. Later, sclerotia and spores were formed on the colonies.

Lorbeer & Tichelaar (1970) and Kritzman & Netzer (1978) developed the selective media to detect onion gray rot caused by *B. allii* from the soil and the seed. Kerssies (1990) modified the composition of Kritzman's medium, and developed a selective medium (Kerssies's medium) for spore trapping of *B. cinerea*. On this medium, the growth of *B. cinerea* colonies was significantly slower and the number of colonies formed was significantly fewer than SBc medium. Bardinell (1989) developed the selective medium (Bardinell's medium) for the fungicide resistant strain diagnosis. When *B. cinerea* was grown on this medium, the color of the medium changed from purple to yellow. This is a positive indication of *B. cinerea* growth, but the areas of discoloration eventually overlapped and accurate calculations of the number of *B. cinerea* colonies were practically not possible.

The comparison of colony formation of *B. cinerea* in SBc and a conventinal selective medium is shown in Table 2. The number of *B. cinerea* colonies in SBc was much more than that of Bardinell's medium or Kerssies's medium. Efficiency for trapping the spores of *B. cinerea* on SBc medium by a spore collection machine was compared with that on the conventional media (Table 3). When the air-borne spores were collected, the average number of *B. cinerea* colonies were 17-36 times more on SBc than those found on Kerssies's medium. Thus SBc medium is considered suitable for the spore collection of *B. cinerea*.

Table 1. Composition of selective medium for *Botrytis cinerea* (SBc)

Basal Medium [1]		*Inhibitors* [2]	
$MgSO_4\ 7H_2O$	0.5g	Streptomycin sulfate	300.0mg
$NaNO_3$	2.0g	Triphenyltin acetate [3]	1.5mg
K_2HPO_4	1.0g	Copper oxide chloride	25.0mg
L-glutamine	2.0g	Rose Bengal	25.0mg
Fructose	10.0g		
Agar	15.0g		
Distilled Water	1000ml		

1) Basal medium is sterilized at 121. for 10minutes.
2) Inhibitors are added to basal medium after cooling to 50..
3) Triphenyltin acetate is dissolved in dimethyl sulfoxide.

Table 2. Comparison of the selective media for *B. cinerea*

Selective medium	*Number of colonies formation of B. cinerea (CFU/petri dish)*
SBc medium	87.5 (83.3%)
Bardinell's medium	6.5 (6.2%)
Kerssies's medium	0.0 (0.0%)
Potato dextrose agar (PDA)	105.0 (100%)

The spores of *B. cinerea* were inoculated on the respective medium, and cultured at 20-25. for 7 days under dark conditions.
Three petri dishes were used in the examination and the average numbers of colonies formed were shown. The values in parentheses indicate percents of colonies formed on each medium compared with those formed on PDA.

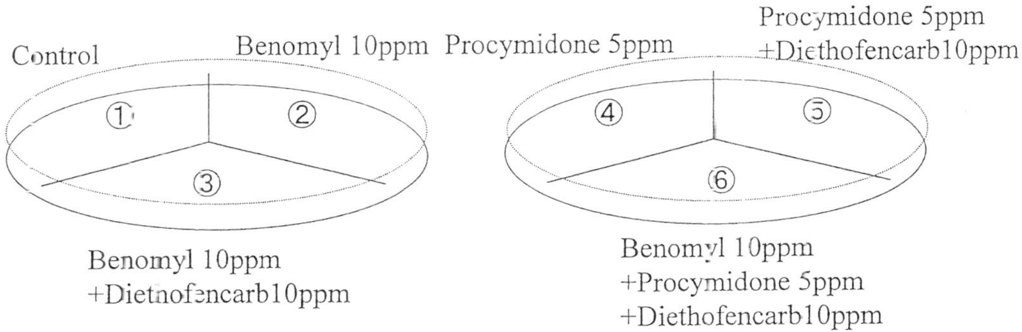

Calculating expression of ratio of each resistant strain

SSR(%) ; (①−②−⑤+⑥)／①×100
SRR(%) ; (⑤−⑥)／①×100
RSS(%) ; (①−④−③+⑤)／①×100
RRS(%) ; (④−⑤)／①×100
RSR(%) ; (③−⑥)／①×100
RRR(%) ; ⑥／①×100

The number of colonies of *B.cinerea* appeared in each section is substituted.

Figure 2. Addition of fungicide to SBc medium for monitoring fungicide resistance and the composition of the examination sections

Table 3. Comparison of the selective media for *B. cinerea* by the spore trap method

Field no.	*Collecting time(min)*	*Number of colonies of B. cinerea on each selective medium (CFU/petri dish)* SBc	*Bardinell's*	*Kerssies's*
Greenhouse1	3	190	nd [1)]	11
	5	612	nd	17
Greenhouse2	3	734	nd	23
Greenhouse3	3	760	nd	31

The air-borne spores of *B.cinerea* were trapped on the media while moving in the greenhouses by using the spore collection machine for 3 to 5 minutes.

1) not detectted.

Table 4. Relative colony formation of each fungicide resistant strain on SBc medium containing fungicides

Fungicide addition [1)]	*Fungicide resistant phenotype (%)* [2)] SSR	*SRR*	*RSS*	*RRS*	*RSR*	*RRR*
No addition	100	100	100	100	100	100
D.10ppm	100	100	0	0	100	100
P.5ppm + D.10ppm	0	100	0	0	0	95
B.10ppm	0	0	100	100	100	100
B.10ppm + P.5ppm	0	0	0	91	0	100
P. 5ppm	0	100	0	100	0	95
B. 10ppm + D. 10ppm	0	0	0	0	85	95
B.10ppm + P.5ppm + D.10ppm	0	0	0	0	0	77

1) B: benomyl, P: procymidone, D: diethofencarb.
2) Fungicide sensitivity to benomyl, procymidone and diethofencarb. S: sensitive, R: resistant.

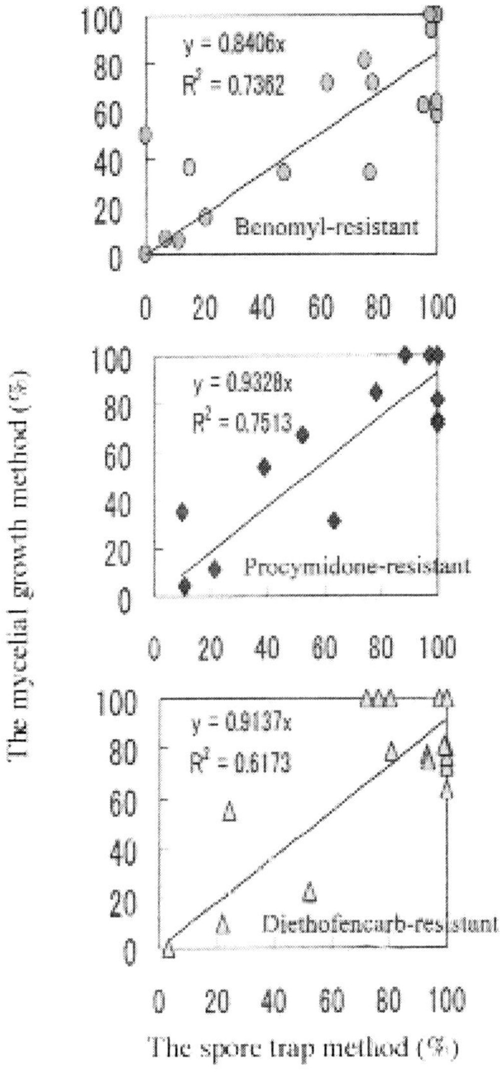

Figure 3. Comparison of the spore trap method and the usual (mycelial growth) method in the detection of fungicide-resistant strains

The spore trap method for monitoring fungicide resistance

Colony formation of fungicide resistant strains of *B. cinerea* on SBc medium containing each fungicide showed 95–100% of the control corresponding to the resistance to benomyl, procymidone and diethofencarb, respectively (Table 4). Thus, each fungicide resistant phenotype was found to reflect its resistance on SBc medium containing the fungicide concerned. The rates of colony formation of isolates, having the phenotype 'benomyl-resistant, procymidone-sensitive, and diethofencarb-resistant' (R, S, R) and 'benomyl-resistant, procymidone-resistant, and diethofencarb-resistant' (R, R, R), were 85 and 77% of the control, respectively. But when these isolates were cultured 7 days or more, they showed almost 100% growth of the control. Therefore, the SBc medium containing fungicides can be used to efficiently distinguish fungicide resistant strains.

The distribution of fungicide-resistant strains monitored by the spore trap method was compared with that of resistant strains monitored by the usual mycelial growth method. When the proportions of benomyl-resistant strains obtained by both the spore trap method and the usual method were plotted, a significant positive correlation ($p=0.001$) was found in the growth rates of strains resistant to benomyl (Fig.3). The inclination of regression is 0.84. Similarly, significant positive correlations were found in the rates of procymidone-resistant strains and diethofencarb-resistant strains. Thus the spore trap method can measure the rate of colony formation of these fungicide-resistant strains with the same accuracy as the usual method.

The monitoring of fungicide resistance by the spore trap method has the following advantages. 1) As the spores are collected while moving in the greenhouse, a uniform sample collection is possible, 2) The time needed for the collection of the samples in a greenhouse is short (3-5 minutes), thus many samples can be obtained in a short time. 3) Even if different types of resistant strains exist together in the same area, growth rate of each fungicide resistant strain can be cleary distinguished on the selective SBc medium.

The distribution of the fungicide resistant strain can be found promptly and accurately by using the Sbc medium containing fungicide(s) and the spore collection machine. Future research should be done to include the testing of other fungicides, such as mepanipyrim, cyprodinil, pyrimethanil and fludioxonil, using this medium and the machine.

References

1. Bardinell, T.R. et al. 1989, Phytopath. 79: 1213(Abstr.).
2. Kerssies, A. 1990, Neth. J. Path. 96: 247-250
3. Kritzman, G. and Netzer, D. 1978, Phytoparasitica, 6: 3-7.
4. Lorbeer, J.W. and Tichelaar, G.M. Phytopath. 60: 1301(Abstr.).
5. Okada, K. 1996, Syokubutsu Boeki, 50: 290-294 (in Japanese).

Author Index

Subject Index

B

C

D

K

L

P

Q

R

S

T

U

V

Bestsellers from ACS Books

The ACS Style Guide: A Manual for Authors and Editors (2nd Edition)
Edited by Janet S. Dodd
470 pp; clothbound ISBN 0–8412–3461–2; paperback ISBN 0–8412–3462–0

Writing the Laboratory Notebook
By Howard M. Kanare
145 pp; clothbound ISBN 0–8412–0906–5; paperback ISBN 0–8412–0933–2

Career Transitions for Chemists
By Dorothy P. Rodmann, Donald D. Bly, Frederick H. Owens, and Anne-Claire Anderson
240 pp; clothbound ISBN 0–8412–3052–8; paperback ISBN 0–8412–3038–2

Chemical Activities (student and teacher editions)
By Christie L. Borgford and Lee R. Summerlin
330 pp; spiralbound ISBN 0–8412–1417–4; teacher edition, ISBN 0–8412–1416–6

Chemical Demonstrations: A Sourcebook for Teachers, Volumes 1 and 2, Second Edition
Volume 1 by Lee R. Summerlin and James L. Ealy, Jr.
198 pp; spiralbound ISBN 0–8412–1481–6
Volume 2 by Lee R. Summerlin, Christie L. Borgford, and Julie B. Ealy
234 pp; spiralbound ISBN 0–8412–1535–9

The Internet: A Guide for Chemists
Edited by Steven M. Bachrach
360 pp; clothbound ISBN 0–8412–3223–7; paperback ISBN 0–8412–3224–5

Laboratory Waste Management: A Guidebook
ACS Task Force on Laboratory Waste Management
250 pp; clothbound ISBN 0–8412–2735–7; paperback ISBN 0–8412–2849–3

Good Laboratory Practice Standards: Applications for Field and Laboratory Studies
Edited by Willa Y. Garner, Maureen S. Barge, and James P. Ussary
571 pp; clothbound ISBN 0–8412–2192–8

For further information contact:
Order Department
Oxford University Press
2001 Evans Road
Cary, NC 27513
Phone: 1 800-445-9714 or 919-677 0977

More Best Sellers from ACS Books

Microwave-Enhanced Chemistry: Fundamentals, Sample Preparation, and Applications
Edited by H. M. (Skip) Kingston and Stephen J. Haswell
800 pp; clothbound ISBN 0–8412–3375–6

Designing Bioactive Molecules: Three-Dimensional Techniques and Applications
Edited by Yvonne Connolly Martin and Peter Willett
352 pp; clothbound ISBN 0–8412–3490–6

Principles of Environmental Toxicology, Second Edition
By Sigmund F. Zakrzewski
352 pp; clothbound ISBN 0–8412–3380–2

Controlled Radical Polymerization
Edited by Krzysztof Matyjaszewski
484 pp; clothbound ISBN 0–8412–3545–7

The Chemistry of Mind-Altering Drugs: History, Pharmacology, and Cultural Context
By Daniel M. Perrine
500 pp; casebound ISBN 0–8412–3253–9

Computational Thermochemistry: Prediction and Estimation of Molecular Thermodynamics
Edited by Karl K. Irikura and David J. Frurip
480 pp; clothbound ISBN 0–8412–3533–3

Organic Coatings for Corrosion Control
Edited by Gordon P. Bierwagen
468 pp; clothbound ISBN 0–8412–3549–X

Polymers in Sensors: Theory and Practice
Edited by Naim Akmal and Arthur M. Usmani
320 pp; clothbound ISBN 0–8412–3550–3

Phytomedicines of Europe: Chemistry and Biological Activity
Edited by Larry D. Lawson and Rudolph Bauer
336 pp; clothbound ISBN 0–8412–3559–7

For further information contact:
Order Department
Oxford University Press
2001 Evans Road
Cary, NC 27513
Phone: 1-800-445-9714 or 919-677-0977